Plants in Human Nutrition

Volume Editor   *Artemis P. Simopoulos*
The Center for Genetics, Nutrition and Health,
Washington, D.C.

26 figures and 65 tables, 1995

KARGER   Basel · Freiburg · Paris · London · New York ·
New Delhi · Bangkok · Singapore · Tokyo · Sydney

....
World Review of Nutrition and Dietetics

Library of Congress Cataloging-in-Publication Data
Plants in human nutrition/volume editor Artemis P. Simopoulos.
(World review of nutrition and dietetics; vol. 77)
Includes bibliographical references and index.
1. Plants, Edible.   2. Nutrition.   3. Plants, Edible–Composition.
I. Simopoulos. Artemis P., 1933– .   II. Series.
[DNLM: 1. Nutritive Value.   2. Plants, Edible.
W1 WO898 v. 77 1995/QU 145 P714 1995]
QP141.A1W59 vol. 77 [QP 144.V44] 612.3 s–dc20 [613.2]
ISBN 3–8055–6101–6 (alk. paper)

Bibliographic Indices. This publication is listed in bibliographic services, including Current Contents® and Index Medicus.

Plants in Human Nutrition

World Review of Nutrition and Dietetics

Vol. 77

Series Editor

Artemis P. Simopoulos,
The Center for Genetics, Nutrition and Health,
Washington, D.C., USA

Advisory Board

Åke Bruce, Sweden
Ji Di Chen, China
Jean-Claude Dillon, France
J.E. Dutra de Oliveira, Brazil
Claudio Galli, Italy
Ghafoorunissa, India
Demetre Labadarios, South Africa
Eleazar Lara-Pantin, Venezuela
Paul J. Nestel, Australia
Konstantin Pavlou, Greece
A. Rérat, France
V. Rogozkin, Russia
Michihiro Sugano, Japan
Naomi Trostler, Israel
Ricardo Uauy-Dagach, Chile

KARGER

Basel · Freiburg · Paris · London · New York ·
New Delhi · Bangkok · Singapore · Tokyo · Sydney

Contents

Preface IX

Microalgae as a Source of ω3 Fatty Acids

Cohen, Z. (Sede-Boker Campus); *Norman, H.A.* (Beltsville, Md.); *Heimer, Y.M.*
(Sede-Boker Campus) 1

Introduction 2
Biosynthesis of Eicosapentaenoic Acid and Docosahexaenoic Acid 6
Environmental and Nutritional Factors Modifying Fatty Acid Content 9
    Growth Temperature 10
    Light 11
    Nitrogen 13
    Silicon 14
Distribution of Eicosapentaenoic Acid and Docosahexaenoic Acid in Microalgae . . 14
ω3 Fatty Acid Production by Phototrophic Algae 15
    Phaeodactylum tricornutum 15
    Nannochloropsis oculata 15
    Monodus subterraneus 16
    Porphyridium cruentum 17
    Isochrysis galbana 19
Selection of Herbicide-Resistant Clones 19
ω3 Fatty Acids Derived from Algal Triglycerides 21
Heterotrophic Eicosapentaenoic Acid and Docosahexaenoic Acid Production 22
Eicosapentaenoic Acid Purification 23
Conclusions: Economic Considerations and Future Directions 24
References 25

Nutritional Value of the Alga Spirulina

Dillon, J.C.; Phan Phuc, A.; Dubacq, J.P. (Paris) 32

History of Spirulina 33
Production 33
Chemical Composition 34
 Composition and Nutritional Value of Proteins 35
 Composition and Nutritional Value of Lipids 38
 Vitamins 41
 Minerals 42
Spirulina as a Human Food 43
Conclusion 44
References 44

Purslane in Human Nutrition and Its Potential for World Agriculture

Simopoulos, A.P. (Washington, D.C.); *Norman, H.A.* (Beltsville, Md.);
Gillaspy, J.E. (Austin, Tex.) 47

Introduction 47
Geographic Distribution and Origin 50
Food and Agricultural Potential of Purslane 51
 Purslane Leaf Content of ω3 Fatty Acids, Antioxidant Vitamins
  (C, E, β-Carotene) and Glutathione 54
  Fatty Acid Content 54
  Antioxidant Content 56
 Profiles of Leaf ω3 Fatty Acids and Antioxidants throughout Plant Development
  in Growth Chamber Conditions 60
 Agricultural Potential of Purslane 64
Medicinal Uses of Purslane 67
Contribution of Purslane to Food Technology: Pectin 69
Conclusions 70
References 71

Sweet Lupins in Human Nutrition

Uauy, R.; Gattas, V.; Yañez, E. (Santiago de Chile) 75

Introduction 75
Nutritional Value of Lupin as a Food 77
Nutritional Value of Lupin Proteins for Humans 80
Significance of Sweet Lupins as Plantfoods for Humans 82
Acknowledgments 86
References 86

Contents

Barley Foods and Their Influence on Cholesterol Metabolism

McIntosh, G.H.; Newman, R.K.; Newman, C.W. (Adelaide/Bozeman, Mont.) . . 89

Introduction 89
Genetic Variation in Barley 91
Composition of Barley Grain 93
    Protein and Lysine 93
    Lipids 94
    Minerals and Vitamins 95
    Carbohydrates 95
Barley and Cholesterol Metabolism 98
Milling and Processing of Barley 103
Conclusion 105
Acknowledgments 105
References 106

The Nopal: A Plant of Manifold Qualities

Muñoz de Chávez, M.; Chávez, A.; Valles, V.; Roldán, J.A. (México City) 109

Introduction 109
    General Background 110
    Environmental Aspects 115
    Cultural Issues 116
Morphological Characteristics 118
Nutrient Content 119
    Proteins and Fats 119
    Vitamins 122
    Minerals 123
    Carbohydrates 123
    Organoleptic Qualities 124
Other Health and Nutritional Characteristics 125
Production and Consumption 127
    Production 127
    Consumption 130
Conclusions 132
References 132

The Corn Tree *(Brosimum alicastrum)*: A Food Source for the Tropics

Ortiz, M.; Azañón, V.; Melgar, M.; Elias, L. (Guatemala City) 135

Introduction 135
Botanical Description 136
    Morphology 137
Uses 138
    Food 138
    Fodder 138
    Wood 139
    Medicinal 139
    Other Uses 139
Distribution 139
Production 139
Nutritive Value 140
    Chemical Composition 141
    Amino Acid Content and Protein Quality of the Corn Tree Seed 142
Development Potential 143
    Tortillas and Bread 143
    Animal Feed 144
    Forestal Resources 144
Conclusions 145
References 145

Hawthorn (Shan Zha) Drink and Its Lowering Effect on Blood Lipid Levels in Humans and Rats

Chen, J.D.; Wu, Y.Z.; Tao, Z.L.; Chen, Z.M.; Liu, X.P. (Beijing) 147

Introduction 147
Materials and Methods 149
    Animal Experiment 149
    Human Study 149
    Indices 149
Results and Analysis of the Animal Experiment 150
Results and Analysis of the Human Study 153
Conclusions 154
References 154

Subject Index 155

....

Preface

The plant kingdom has supplied us with many therapeutic ingredients: colchicine for gout, vinca alkaloids for leukemia, liquorice for peptic ulcers, salicylates for pain and fever, curare for relaxant anesthesia, quinine for malaria, etc. This volume, *Plants in Human Nutrition*, reflects research advances and the recognition by the biomedical, pharmaceutical, and the agricultural communities that plant foods not only represent the major source of nutrients for humans, but also contain 'protective factors' against chronic diseases, coronary heart disease, diabetes and cancer. For this reason, the volume includes the regulation of biosynthetic pathways of polyunsaturated fatty acids in algae as a prelude to genetic manipulation for increased production – to descriptions of nutrient composition of selected plants from around the world that may explain their medicinal use through the ages in different cultures – and finally to the current use of plant foods in clinical studies.

In 1989 the Diet and Health Committee of the US National Research Council recommended that Americans should increase their consumption of fruits and vegetables to five or more one-half cup servings per day, especially green and yellow vegetables and citrus fruits. The basis of their recommendation was the association of diets high in plant foods (whole grain cereals, legumes, fruits and vegetables) with a lower occurrence of coronary heart disease and cancers of the lung, esophagus and stomach. Although the biochemical mechanisms for this association are not fully understood, the committee felt that it could be largely accounted for by the following nutritional characteristics of the diet: low total and saturated fat and cholesterol, high complex carbohydrate content (high fiber), and high content of vitamins A and C. Since then, additional studies indicate that α-tocopherol may prevent heart disease in humans, and ω3 fatty acids may prevent heart disease in humans and cancer in animals. Green leafy vegetables are good sources of α-tocopherol, ascorbate, β-carotene, and α-linolenic acid (18:3ω3). The latter elongates and desaturates to eicosapentaenoic acid (EPA;

20:5ω3) and docosahexaenoic acid (DHA; 22:6ω3) in the human body. EPA and DHA are found in the oils of fish and are also produced by some algae.

In addition to the types of food, recent studies show that food variety is also essential to health, and raise the intriguing possibility that a narrow range of foods may antedate diabetes or even predispose to its expression. Of the 1,300 known food plants, fewer than 20 are currently providing most of our food needs, making us heavily dependent on a few crops to provide food, yet plants have influenced the evolution of human beings. Paleolithic humans ate a greater variety of wild plants and fruits that were richer in vitamins and minerals than today's limited cultivated varieties. The plants included in this volume (algae, spirulina, purslane, corn tree, nopal, lupin, barley and hawthorn) are of enormous agricultural, medical and pharmaceutical potential. Furthermore, they have been part of man's diet for thousands of years.

In selecting the topics for this volume, we included plants that had some or all of the following characteristics: they are excellent sources of ω3 fatty acids; are rich sources of antioxidant vitamins (α-tocopherol, ascorbate, β-carotene); contain high amounts of glutathione; are rich in fiber; are high in protein content, and can grow in arid climates. Therefore these plants should contribute to the improvement of the food supply in many parts of the world.

The importance of dietary supply of ω3 fatty acids relates to their essentiality for normal growth and development of infants and on their positive effects on various diseases and disorders such as coronary artery disease, thrombosis, hypertension, cancer of the colon, arthritis, and other autoimmune disorders. Dietary ω3 fatty acids have been shown to have hypolipidemic, antithrombotic, and anti-inflammatory properties. The sole commercial source of longer-chain ω3 fatty acids (EPA and DHA) is from marine fish oil. Because of the great demand that is projected worldwide, alternative sources of ω3 fatty acids are being investigated in the first paper by Cohen et al. on 'Microalgae as a Source of Omega-3 Fatty Acids'. This review summarizes the results of studies designed to identify algal species which are the best potential sources of EPA and DHA under phototrophic conditions, and evaluates current approaches towards increasing these compounds in algae. Available knowledge of the biosynthetic pathways for polyunsaturated fatty acids in algae is also discussed since future attempts to regulate EPA and DHA pathways are a prelude to genetic manipulation. The unique composition of algal oils already offers distinct advantages for large-scale and efficient purification of EPA and other polyunsaturated fatty acids relative to other sources. Microalgae as a source of ω3 fatty acids is an area of immense interest in the US, Japan and other parts of the world. The authors discuss economic considerations and future directions.

The second paper by Dillon et al. is on the 'Nutritional Value of the Alga Spirulina'. The authors provide information on the historical, agricultural, and nutritional aspects of spirulina. Spirulina represents one of the richest sources of protein of plant origin (up to 70% of dry weight). It is a good source of the B vitamins (niacin), and is high in carotene. It is the richest nonanimal source of cobalamine (vitamin B_{12}). One of the spirulinas, *Spirulina platensis* is unique in that it contains exclusively γ-linolenic acid (18:3ω6) and it is the most concentrated source of γ-linolenic acid in the vegetable kingdom (1% of dry weight): 25–30% of fatty acids are 18:3ω6, whereas other vegetable sources such as evening primrose or black currant berry contain only 10–15%. It is a good source of iron as iron sulfate. It is safe as human food. Today, while commercial farms grow spirulina as a health food, village scale projects in Africa, Asia and South America produce spirulina for local consumption and help fight protein and vitamin A malnutrition.

The third paper is on 'Purslane in Human Nutrition and Its Potential for World Agriculture' by Simopoulos et al. Information is provided on the historical, agricultural, nutritional, medicinal and pharmacological aspects of purslane. Purslane has the widest geographic distribution of any other edible weed or plant, and has enormous agricultural potential since it has C4 metabolism, and can grow in arid lands. Purslane is thought to be one of man's earlier vegetables. It has been used for human food and medicine, and for animal feed throughout the world for thousands of years. Being such a cosmopolitan plant, purslane has similar medicinal uses around the world and has been an important food for Australian Aborigines, Indians, Africans, American Indians, the Caribs and the people in and around the Mediterranean basin. Purslane is a richer source of α-linolenic acid than any other green leafy vegetable. 100 g of purslane leaves provide 400 mg of α-linolenic acid, 27 mg of ascorbic acid, 12 mg of α-tocopherol, 15 mg of glutathione, 2 mg of β-carotene, and have a high content of pectin (25 g%, dry weight), which along with the high content of α-linolenic acid, α-tocopherol, ascorbic acid, glutathione, and β-carotene, could account for its medicinal uses. In a recent article, purslane was considered a 'power food'.

The next paper, by Uauy et al., is on 'Sweet Lupins in Human Nutrition'. Lupin has been cultivated for six millennia in the New World and for over 3,000 years in the Old World. Egyptians, Greeks and Romans used *Lupin albus* as a grain and as a soil enricher (nitrogen fixing). It grows in poor soils and in adverse climatic conditions. Research on lupins has generated several low alkaloid cultivars. The low-saturated, high-monounsaturated, and balanced ω3/ω6 fatty acids are similar to those of soya oil and likely to be comparable in terms of health benefits. The protein content of whole lupin seeds is as high or higher than that of soybeans. Lupin protein, as is true for all legumes, has a low sulfur amino acid content. The main advantages of lupin relative to other legumes used in human nutrition relate to its high protein content. Although deficient in sulfur amino acids, lupin protein is complementary to cereal proteins, thus the mix will be of higher biological value. Lupins have twice as much protein as beans, chick peas, lentils, and other legumes. Lupin flour can be defatted: this concentrated protein product (60–70% protein) blended with processed cereal flours and milk (lupin-wheat-milk) can be used for beverages or milk imitation products. These are less costly than milk and may use local agricultural products rather than imported dairy materials. Most developing countries have limited milk production capacities. Lupin protein isolates have also been prepared by alkaline extraction and acid precipitation. The testing of functional properties revealed better solubility than soya isolates and similar emulsion capacity. It is possible that lupin may replace soybean protein isolates in many of their uses. Lupins can be cultivated in extreme geographic conditions such as in Iceland, Alaska or Patagonia at the southern tip of the American continent. The nitrogen fixation properties of lupins are greater than those of other legumes, furthermore the citric exudates produced by the roots mobilize phosphorus and other minerals fixed in the soil. These naturally protected lupins may open the way to ecologically gentler and kinder agricultural practices since less fertilizers and pesticides are required. Based on these facts, nitrogen fixation, phosphorus mobilization, and climatic tolerance, lupins offer an excellent alternative for sustainable agricultural production with a positive impact on the environment.

Barley is a good source of fiber and has hypocholesterolemic properties. McIntosh et al. present their extensive studies in their paper on 'Barley Foods and Their Influence on Cholesterol Metabolism'. Barley grain has served mankind well through thousands of years, as a good source of energy, dietary fiber, protein, and other essential nutrients. In the last century, barley has been replaced by wheat, rice and maize. Barley is possibly the earliest cereal grain in recorded history. It is mentioned several times in the *Bible.* Archaeologists have found carbonized grains of barley in Southern Egypt nearly 9,000 years old. Barley is the fourth most important cereal in the world in terms of world production (12% of total cereal production). With the increasing interest in the use of barley as a food grain, its wide genetic diversity provides the capability of meeting a variety of nutritional, functional and food ingredient needs while maintaining acceptable agronomic characteristics. Studies on the chemical composition of barley have previously been conducted mainly by agronomists, animal scientists, and brewing technologists, rather than by food scientists. The authors review human studies that show indisputable evidence that the β-glucan component of soluble dietary fiber in barley and oats is a useful agent for cholesterol control in hypercholesterolemic individuals. Nonsaponifiable components (β-carotene, tocotrienol) influence cholesterol metabolism by inhibiting HMG-CoA reductase enzyme. Also 6% of its oil is α-linolenic acid. Barley provides a concentrated source of dietary fiber for the food industry.

De Chavez et al. describe the historical and environmental aspects as well as the cultural influence of nopal on the people of Mexico throughout history in their paper on 'The Nopal: A Plant with Manifold Qualities'. Nopal is used as food by humans and animals. Humans eat the fruit (prickly pear) as well as the leaves (called 'cladodes'). The leaves are eaten as vegetables, whereas the fruit is peeled and eaten raw, or ground to produce cheese, or fermented for beer (called 'colanche'). The cladodes are considered to have medicinal properties. The authors present information suggesting that the cladodes cooked as vegetables reduce serum cholesterol levels, regulate blood sugar and control gastric acidity, and emphasize the need for further studies, including clinical trials, to specifically define its use (the whole fruit or the cladodes, or special extract) for its medicinal properties and the mechanisms involved. Besides the outstanding nutritional, medicinal and ceremonial aspects, the authors provide information on the many other uses of the nopal plant to control soil erosion, as a fence to separate land holdings and for construction. The mucilaginous material that is contained in the leaves of the nopal has been used to glue bricks together in building houses. The old cladodes are used as fuel to cook the traditional tortillas and beans. Nopal hosts the cochineal insect which reproduces on the fleshy leaves and produces the scarlet red dye since pre-Hispanic times. Nopal has a wide distribution. Some species, such as *Opuntia stricta* grow at sea level, while other such as *Opuntia streptacantha* (giant cactus), grow at 2,700 m or more above sea level. Today, only three countries produce nopal for commercial purposes: Mexico, Chile and Italy (mostly Sicily). Only Mexico produces cladodes as a vegetable, and a new thornless variety is exported to the US. The total number of species is about 258 with 110 being well defined. All species grow wild and only 8 of them are used for human consumption. *Opuntia ficus indica* is most commonly cultivated for its fruits. Nopal is a rich source of fiber. The cladodes are rich in lysine, methionine and tryptophan, the amino acids in which corn is poor. When the prickly pear is made into cheese, then the concentration of protein increases due to the ground seeds that the fruit contains. The cladodes are rich in β-carotene and the fruit in vitamin C. The nopal is rich in calcium, magnesium and potassium. At present, the use of nopal (prickley pear and cladodes) is becoming popular as an antidiabetic agent in Mexico and is imported by Japan and Germany for this purpose also.

In the next paper on 'The Corn Tree *(Brosimun alicastrum):* A Food Source for the Tropics', Ortiz et al. describe the various uses of the corn tree: as food, fodder, wood and medicine. Information is provided on the corn tree's distribution, production, its nutritive value and its developmental potential for human consumption in making tortillas and bread; for animal feed, and forestal resources. The corn tree and corn were the main food sources of the Mayas. The seed was used as food and the trees are found near the sites of Mayan ruins, a relic of ancient horticulture. The authors provide information on the enormous potential of the plant for the tropics, as a food, medicine, fodder, wood and as an ornamental plant also contributing to environmental quality of life. The seed is carbohydrate rich, is of high caloric content, and has a protein content of 7.7–8.9%. Thus the seed of the corn tree compares favorably with wheat, corn, and rice whose average protein values are 9.3, 9.8 and 7.2%, respectively. The authors point out that in the wet tropics where corn is not an efficient crop, the corn tree is a very good alternative. The authors state that the Mayas must have known the importance of the plant's nutritive potential and its importance in the conservation of the forests, and regret that in many tropical regions tubers, e.g. yams and potatoes, which are poor sources of proteins, 1.3 and 1.8% respectively, have replaced cereals. Therefore, the use of corn tree as a partial substitute for corn would satisfy the nutritional requirements of the populations living in the wet tropics.

The last paper is on 'Hawthorn (Shan Zha) Drink and Its Effect on Lowering Blood Lipid Levels of Humans and Rats' by Chen et al. Hawthorn has been used both as food and for its medicinal properties which include the treatment of heart disease to reduce hypercholesterolemia, angina pectoris and hypertension. Dr. Chen describes their studies using the whole berry to make a drink, rather than attempting to isolate the various components in carrying out their studies on rodents and humans. Whereas the modern pharmacopoeia is based on the isolation of the active components from plants or herbs, and the eventual synthesis of drugs for the treatment of diseases, a renewed approach, that of using the whole fruit or plant for both the prevention and treatment of chronic diseases, may be emerging, as shown by these studies.

Plants in Human Nutrition should be of interest to those involved in food production, industrial and agricultural development, and sustainable agriculture, including scientists who are students of human evolution and development. Specifically, botanists, experimental biologists, agronomists, food technologists, nutritionists, pharmacologists, physicians, economists, policy makers, and anthropologists will discover their collective contribution in furthering human health and sustainable agriculture, and having a positive impact on the environment.

Artemis P. Simopoulos

Simopoulos AP (ed): Plants in Human Nutrition.
World Rev Nutr Diet. Basel, Karger, 1995, vol 77, pp 1–31

....

Microalgae as a Source of ω3 Fatty Acids[1]

Zvi Cohen[a], *Helen A. Norman*[b], *Yair M. Heimer*[c]

[a] The Laboratory for Microalgal Biotechnology, Sede-Boker Campus, Israel;
[b] Weed Science Laboratory, United States Department of Agriculture, Agricultural
Research Service, Beltsville, Md., USA;
[c] Desert Agrobiology Research Unit, Jacob Blaustein Institute for Desert Research,
Ben-Gurion University of the Negev, Sede-Boker Campus, Israel

Contents

Introduction 2
Biosynthesis of Eicosapentaenoic Acid and Docosahexaenoic Acid 6
Environmental and Nutritional Factors Modifying Fatty Acid Content 9
    Growth Temperature 10
    Light 11
    Nitrogen 13
    Silicon 14
Distribution of Eicosapentaenoic Acid and Docosahexaenoic Acid in Microalgae . . 14
ω3 Fatty Acid Production by Phototrophic Algae 15
    Phaeodactylum tricornutum 15
    Nannochloropsis oculata 15
    Monodus subterraneus 16
    Porphyridium cruentum 17
    Isochrysis galbana 19
Selection of Herbicide-Resistant Clones 19
ω3 Fatty Acids Derived from Algal Triglycerides 21
Heterotrophic Eicosapentaenoic Acid and Docosahexaenoic Acid Production 22
Eicosapentaenoic Acid Purification 23
Conclusions: Economic Considerations and Future Directions 24
References 25

[1] Contribution No. 64 from the Laboratory for Microalgal Biotechnology, Jacob Blaustein Institute for Desert Research.

Introduction

Polyunsaturated fatty acids (PUFAs) consist of two families known as ω3 and ω6 fatty acids, each with methylene interrupted double bonds. This nomenclature is based on double-bond location in the fatty acid molecules (fig. 1). In the ω3 family (or n–3 fatty acids), the first double bond is located 3 carbon atoms from the terminal methyl group (the ω- carbon), whereas in ω6 (or n–6) fatty acids, the first double bond is located between the 6th and 7th carbon atoms. The parent ω3 fatty acid is α-linolenic acid (LNA; 18:3ω3) and the parent ω6 fatty acid is linoleic acid (LA; 18:2ω6). LNA is found in green leafy vegetables and in linseed and rapeseed oils, whereas LA is found predominantly in the seeds of all plants except for palm, cocoa and coconut. In humans, LNA is metabolized to eicosapentaenoic acid (EPA) and docosahexaenoic acid (DHA), although at a rather low rate, and LA is metabolized to arachidonic acid (AA). EPA and DHA are ω3 PUFAs commonly found in marine fish and mammals, and phytoplankton. EPA contains 20 carbon atoms with 5 double bonds (20:5ω3), whereas DHA contains 22 carbon atoms with 6 double bonds (22:6ω3). AA (20:4ω6) also found in these sources, belongs to the ω6 family of fatty acids (fig. 1).

In typical Western diets the ω6 fatty acid LA is present in high quantities in vegetable oils, and AA is consumed in meats, eggs and dairy products. Therefore Western diets are typically high in ω6 fatty acids relative to ω3 fatty acids. Dyerberg et al. [1], studying the very low incidence of coronary vascular disease in Greenland Eskimos, were the first to attribute this phenomenon to differences in nutrition, and particularly to the high level of ω3 fatty acids in the Eskimo diet, originating in marine fish and mammals. These and subsequent studies led to the conclusion that consumption of food rich in ω3 PUFAs results in higher ω3:ω6 fatty acid ratios in cell membranes and plasma lipids, the assumed reason for a lower incidence of coronary disease.

The importance of dietary ω3 fatty acids relates to their positive effects on various diseases and medical disorders, including cardiovascular problems such as atherosclerosis [2, 3], thrombosis [4], hypertriglyceridemia [5], and high blood pressure [6], as well as several cancers such as breast [7], prostate and colon cancers [8]. A third area includes treatment of various tissue inflammations such as asthma [9], arthritis [10], lupus erythematosus [11], psoriasis [12] and nephritis [13].

The most plausible explanation for the cardiovascular effects of ω3 fatty acids results from their ability to alter the balance of prostaglandin and leukotriene eicosanoids which mediate inflammation and immune responses [14]. It has also been shown that elevated ω3 fatty acids decrease blood viscosity and increase membrane fluidity and deformability [15]. The balance of ω3 and ω6 fatty acids is crucial since some of the eicosanoids derived from EPA and AA are

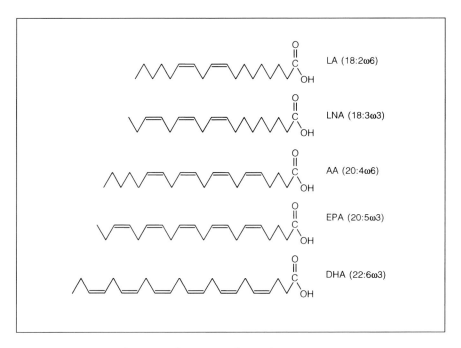

Fig. 1. Structural formulae of ω3 and ω6 fatty acids.

antagonistic [16], and increased EPA levels can inhibit the formation of eicosa-noids derived from AA by competing for the same cyclooxygenase enzyme [17]. Considerations such as these are important when evaluating dietary supplementa-tions.

Independent studies have suggested separate roles for EPA and DHA [18, 19]. The latter is found mainly in brain and retinal tissues, and continued devel-opment of these tissues in the human embryo and newborn requires a constant supply of DHA. Cow's milk does not contain DHA, and current infant formulae are devoid of it altogether, therefore some attention has been focussed on devel-oping DHA supplements that would also include AA. Studies have shown that formula-fed babies had lower DHA levels in plasma phospholipids (PLs) and red cell membranes compared with breast-fed babies, which was correlated with a retarded rate of visual development [20, 21]. The latter could be alleviated by supplementing formulae with DHA [22]. AA is also essential for infant develop-ment [22]; a recent patent for an improved infant formula includes both DHA and AA [23].

The medical importance of ω3 fatty acids has led to increased interest in identification and efficient exploitation of natural sources of these compounds. At present, the sole commercial source of EPA and DHA is marine fish oil. However, satisfactory exploitation of this source is hampered by several drawbacks such as variations in oil quality, the occurrence of fatty acids of antagonistic properties such as AA, and the fact that fish do not synthesize EPA efficiently. The exact ω3 fatty acid composition of fish oil depends upon species, season, and geographic location of harvest and on the availability and types of the primary food chain, namely marine microorganisms [24].

On a global basis, about one million tons of fish oil is produced annually which contains approximately 100,000–250,000 tons of EPA and DHA [25]. Most fish oil produced today is hydrogenated and incorporated into margarine and shortening. Even if they were used solely for extraction of EPA and DHA, the available supply would be sufficient for only 50–100 million people assuming a daily supplement of 5 g. Furthermore, due to the low EPA content of fish oil, and the costliness of purification, the need to comply with the dose recommended for prevention of cardiovascular ailments would require a 5- to 10-fold increase in oil intake. Overall, the supply of ω3 fatty acids solely from fish is unlikely to meet future demands, and it is therefore necessary to develop alternative sources.

Microalgae comprise an extremely diverse group of microorganisms, including marine phytoplankton, which provide a primary source in the oceanic food chain. Most microalgae are photoautotrophic, requiring only light, carbon dioxide and other minerals for growth. Some marine microalgae and a few species of marine bacteria are capable of de novo synthesis of EPA and/or DHA [26, 27], and these ω3 fatty acids present in fish oil can therefore be derived from plankton in the fish diet [24, 28]. It has been confirmed that salmon which have been raised in captivity and fed only soy-based chow contain very little EPA and DHA [29]. The limitations of fish oil have already initiated several studies aimed at identification and development of specific microalgae, marine bacteria [30], and fungi [31] for possible future commercial production of fatty acids. Very recently, several Japanese laboratories have also reported the production of AA and EPA by species of *Mortierella* [30, 32, 33], however, large-scale cultivation of this organism is not yet feasible.

Microalgae, comprising at least 30,000 species [34], represent in effect an untapped natural resource. Of the limited number of species so far analyzed for their lipid composition, an appreciable proportion displayed fatty acid profiles that differed from the typical composition of higher plants in containing significant levels of long-chain (over 18 carbon atoms) PUFAs (table 1) [35–45]. Algal oils may provide a cleaner and more concentrated source of EPA and DHA than fish. Their quality can be controlled, and thus they could be utilized for future pharmaceutical-grade production of these compounds. In addition, algal biomass could be utilized in aquaculture to enhance the ω3 PUFA content of fish intended for human consumption or oil production. Commercial cultivation of microalgae

Table 1. Fatty acid composition of EPA- and DHA-rich microalgae

	$C_{14}+C_{16}$ fatty acids	18:0	18:1	18:2	18:3	18:4	20:3	20:4 (AA)	20:5 (EPA)	22:6 (DHA)	Reference
Diatoms											
Asterionella japonica	51.0		3.0			1.0	10.0	20.0			35
Bidulphia sinensis	71.3	1.1	1.0					24.2			36
Chaetoceros septentrionale	51.0		3.0			1.0	1.0	21.0			35
Lauderia borealis	56.3		1.8	1.1			0.3	1.3	30.3		37
Cylindrotheca	54.3	1.0	5.3	2.9				6.4	24.4		38
Navicula biskanteri	56.1	2.7	7.2		2.4				26.7		38
Stauroneis amphioxys	56.1	1.4	2.2	1.1	2.8	2.8		2.9	26.1	1.9	39
Cyclotella cryptica	69.0	2.0	3.0	2.0	1.0				21.0		35
Navicula pelliculosa	67.0	1.0	2.0	2.0	1.0				26.0		35
Bidulphia aurtia	74.0								26.0		40
Nitzschia alba	41.0		20.0	4.0	1.0				29.0		40
Nitzschia closterium	50.0		4.0	4.0					45.0		40
Skeletonema costatum	58.0		2.0	2.0		1.0		4.0	30.0		35
Chrysophyceae											
Pseudopedinella sp.	61.0		3.0	2.0			1.0	1.0	27.0		35
Cricosphaera carterae	64.0		3.0	3.0			2.0	2.0	20.0		35
Cricosphaera elongata	59.0		2.0	3.0		1.0		2.0	28.0		12
Isochrysis	31.5		20.6	4.0		20.8				17.6	35
Eustigmatophyceae											
Monodus subterraneus	46.0		3.7	2.7	1.1			5.1	39.2		*1
Nannochloropsis	36.7		1.4	1.7			1.2	3.4	44.7		41
Olisthodiscus sp.	37.1		1.1	1.0	7.1	7.9		2.2	21.8	3.0	42
Rhodophyceae											
Rhodella violacea 115.79	37.2	1.0	6.3	1.6	0.5	0.7	1.0	18.9	32.9		*
Porphyridium cruentum 1380.1d	34.1			4.9	0.6	0.6	0.4	14.7	44.1		43
Prymnesiophyceae											
Pavlova salina	35.6			1.8	2.0	15.2			28.2	10.9	36
Prasinophyceae											
Heteromastix rotunda	42.9		2.3	2.8	3.7	9.3			27.9		44
Dinophyceae											
Amphidinium carteri	17.0	2.0	2.0	1.0	3.0	15.0			20.0	22.0	35
Ceratium furca	35.1		4.8	3.1		1.0			7.4	20.7	45
Cochlodinium heteroloblatum	37.7	1.0	6.4	2.4		1.4			11.4	28.0	45
Crypthecodinium cohnii	39.0	1.0	14.0							30.0	37
Glenodinium sp.	33.0	3.0	5.0	5.0	6.0	23.0			2.0	19.0	37
Gonyaulax polyedra	40.0		3.0	2.0	3.0	14.0			14.0	23.0	35
Gonyaulax catenella	23.0	1.2	2.7	1.9	1.8	7.3			11.2	33.9	45
Gyrodinium cohnii	40.0		14.0							30.0	35
Peridinium triquetrum	40.7	1.2	4.8	4.8		4.3			5.1	21.3	45
Prorocentrum minimum	26.7	1.3	4.3	5.4		4.2			5.3	24.5	45

[1] Cohen, unpublished.

has so far been attempted in only a few species but never for the production of PUFAs, with the main limitation currently being the high cost of biomass production which demands unique technologies not yet fully developed.

The EPA and DHA content of microalgae so far tested has been found to be unsuitably low for commercial exploitation. However, it is believed that the combined use of physiological, biochemical and genetic approaches will result in an enhanced productivity of these fatty acids by algae. This achievement together with a reduction in the cost of biomass production should lead to economic viability. An initial crucial factor will be the enhancement of the content of the target fatty acids in algal cells. In this chapter, we will review the various efforts taken to reach this goal.

Biosynthesis of Eicosapentaenoic Acid and Docosahexaenoic Acid

Elucidation of biosynthetic pathways will provide the basis for future genetic engineering approaches aimed at development of improved algal strains with enhanced levels of EPA and DHA. Unfortunately, little is known about the metabolic pathways of PUFAs in algae, and what is known appears to be species specific.

Precursor fatty acids include LA and LNA. Two major pathways for lipid-linked fatty acid desaturation to form LNA in higher plants and microalgae have been proposed [46–49]. These have been designated as the 'prokaryotic' and 'eukaryotic' pathways, resulting in the biosynthesis of the major polyunsaturated molecular species of monogalactosyldiacylglycerol (MGDG), 18:3/16:3 and 18:3/18:3 MGDG, respectively. By the prokaryotic scheme, MGDG is first synthesized as the 18:1/16:0 species within the chloroplast, and is subsequently desaturated at both positions in situ to form 18:3/16:3 MGDG. In the second (eukaryotic) pathway of 18:3ω3 synthesis, 18:1-CoA, synthesized in the chloroplast, is exported across the chloroplast envelope into the endoplasmic reticulum, where it is employed in the synthesis of 18:1-phosphatidylcholine (PC). After desaturation to 18:2-PC, the PL is either further desaturated to 18:3-PC or transported back to the chloroplast for galactosylation of the diacylglycerol moiety yielding MGDG which is finally desaturated to the 18:3/18:3 species. These sequences result in distinctive lipids: prokaryotic lipids possess C_{16} fatty acids at the *sn*-2 position, and both C_{16} and C_{18} at the *sn*-1 position of glycerolipids, while eukaryotic lipids possess C_{18} fatty acids at the *sn*-2 and both C_{16} and C_{18} fatty acids at the *sn*-1 position.

Red algae can contain both EPA and AA. Araki et al. [50] have proposed that the red macroalgae can be divided into four types according to the predominance of these PUFAs in various lipid classes. Type I consists of species such as *Graci-*

laria verrucosa containing high levels (51–61% of fatty acids) of AA, and very little if any EPA. AA is found in all lipid classes but mainly in MGDG and PC. In type II species (e.g. *Porphyra yezoensis*), EPA is the main C_{20} fatty acid (30–39% of fatty acids), but they also contain significant levels of AA (10–23%), particularly in PC and triglycerides (TGs). EPA was primarily found in MGDG. The red microalga *Porphyridium cruentum* also falls into the type II category [51]. The third group consists of species containing both EPA and AA as the major PUFAs, such as *Chondrus ocellatus*. Species of the fourth type contain C_{16} unsaturated fatty acids (16:1 and 16:3) in addition to EPA and AA (e.g. *Meristotheca papulosa*).

In the diatom *Phaeodactylum tricornutum*, EPA was found primarily in MGDG (34% of fatty acids) and PC (31%) [52]. It can be deduced that the major molecular species of MGDG in this alga are 20:5/16:3, 20:5/16:1 and 20:5/20:5. This is analogous to higher plants where 18:3/16:3 molecular species are formed via the prokaryotic pathway and 18:3/18:3 species in the eukaryotic pathway. It is also in keeping with the suggestion that in many algae, EPA fulfills the role of LNA in higher plants [53]. Since EPA is found in molecular positions equivalent to that of LNA, it can be proposed that EPA is formed by further elongation and desaturation of LNA-bound galactolipid (GL) molecular species, although the exact biosynthetic pathways have not yet been characterized. Algal strains which predominantly synthesize GL via the eukaryotic pathway are advantageous for EPA production, since their GL will carry the 20:5/20:5 MGDG molecular species which contains twice as much EPA as the 20:5/16:0 species originating from the prokaryotic pathway.

Arao and Yamada [54] concluded that EPA is restricted to the *sn*-1 position in MGDG in algae belonging to *Phaeophyta* (brown algae), *Bacillariophyta* and *Rapidophyta,* while in the *Rhodophyta* (red algae), EPA was found also in the *sn*-2 position. The occurrence of EPA at the *sn*-1 position and that of 18:2 at the *sn*-2 position in GL of brown algae led them to suggest that 20:4/18:1 GL is further desaturated to 20:5/18:2-GL in these species.

Compositional and metabolic studies with *P. tricornutum* [55] indicated the possible biosynthetic route of EPA shown in figure 2. By this scheme, 18:2 desaturation to 18:3ω3 and 18:4ω3 is followed by elongation to 20:4ω3, which is finally desaturated to EPA. Support for this sequence came from studies of fatty acid metabolism in *Nannochloropsis* sp. [41]. It has been proposed that AA biosynthesis can proceed via two pathways leading from 18:2ω6 to 20:3ω6 (i.e., 18:2→18:3ω6→20:3ω6 and 18:2→20:2ω6→20:3ω6), with the first pathway predominating in several different algal classes (as well as animals) and the second predominating in the *Euglenophyceae* [44].

Using radioactive-labeled fatty acids, Nichols and Appleby [56] have shown that in *P. cruentum,* 18:2 is converted to AA via 18:3ω6 (γ-linolenic acid; GLA) and 20:3ω6, while in *Ochromonas danica* GLA is converted to AA also via

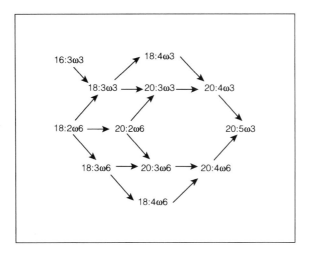

Fig. 2. Possible elongation and desaturation pathways of PUFA biosynthesis in microalgae.

18:4ω6. In *Euglena gracilis,* however, 18:2 is first elongated to 20:2ω6, which is subsequently desaturated to 20:3ω6 and AA.

Chemical inhibitors are useful in identifying precursors in metabolic pathways. The substituted pyridazinone compound, 4-chloro-5-dimethylamino-2-phenyl-3(2H)-pyridazinone (SAN 9785; BASF 13-338), has been widely shown to selectively reduce LNA levels in plant tissues [57]; comparison of the effects of SAN 9785 on leaf lipid metabolism in wild-type and mutant *Arabidopsis* suggested that chloroplast LA desaturase activity forming polyunsaturated MGDG is selectively inhibited [58]. In *P. cruentum,* this herbicide caused a decrease in the level of EPA and an increase in LA [59]. In *Monodus subterraneus,* SAN 9785 reduced growth and decreased fatty acid content, but surprisingly affected an increase in the proportion of EPA, which in both species is primarily located in GL [51]. This contradictory effect of SAN 9785 in *M. subterraneus* suggests the existence of an alternative pathway for EPA biosynthesis in this alga. According to this scheme, elongation of 18:2ω6 to 20:2ω6 would precede desaturation of the latter to 20:3ω3 (fig. 2) which would then be further desaturated to form 20:5ω3. Based on the data presented here, our hypothesis is that these desaturations of C_{20} fatty acids are much less sensitive to the inhibitor. Another possibility is that in this alga the desaturation of 18:2ω6 to 18:3ω3 also takes place on extrachloroplastic PC as part of a eukaryotic pathway, and that this desaturation is less sensitive to SAN 9785 inhibition, as was suggested for higher plant leaf tissue [58, 60]. Similar results were reported for *Nannochloropsis oculata* by Henderson et al.

[61]. Additional data supporting the suggested role of 18:3ω3 and 18:4ω3 as EPA precursors, and indicating the existence of an alternative pathway for EPA, which is SAN 9785 insensitive, were also obtained with *Chroomonas salina* [61, 62].

Various reports indicate that EPA biosynthesis in diatoms may proceed via a different pathway. Nichols and Appleby [56] suggested that palmitoleic acid (16:1) may be elongated to 18:1, which is desaturated to 18:2 and further elongated to 20:2ω3. It was proposed [63–66] that at least three pathways of PUFA biosynthesis exist in microalgae and confirmed that the initial steps of EPA biosynthesis in the diatom MK8908 are routed through 16:1 (i.e. 16:0→16:1→ 18:1), and are then followed by the predictable conversion to 18:2 and 18:3. Elongation and desaturation of 18:2-CoA in a cell-free system was also demonstrated, indicating the possible existence of a thioester-linked desaturation mechanism.

Some fungi have been shown to convert AA to EPA [67] and it has been suggested that this mechanism may also exist in algae, especially in those containing high levels of both fatty acids, such as *P. cruentum.*

Although EPA biosynthesis involves polar lipid substrates, there are apparently some transacylation mechanisms that are responsible for the transfer of EPA between polar lipids and TGs [68]. Their existence was indicated by the decrease in polar lipid EPA during TG accumulation in *N. oculata.* Sukenik and Livne [69] provided correlative evidence to the effect that the enzyme acetyl-CoA carboxylase has a regulatory role in lipid synthesis and accumulation in unicellular algae.

DHA is almost exclusively the only PUFA present in the marine dinoflagellate *Crypthecodinium cohnii* (table 1). Since this organism is devoid of a chloroplast and hence nonphotosynthetic, it can aid in the characterization of the enzymes involved in the production of DHA [70]. PC, the major PL of *C. cohnii,* has been implicated as a substrate for desaturation steps in the biosynthesis of DHA in this alga [71].

Environmental and Nutritional Factors Modifying Fatty Acid Content

The fatty acid composition of algae can be modulated by growth conditions, although as discussed below, the changes may be species specific. Alteration of culture conditions (environmental and nutritional) may therefore provide a means for optimization of algal PUFA content for commercial production.

When growth is retarded in response to any limiting factor (e.g. low light intensity, nutrient depletion, or nonoptimal pH, temperature or salinity), synthesis of lipids and carbohydrates may be enhanced at the expense of protein synthesis. Generally, the newly synthesized lipid will largely consist of TG reserves, and since the fatty acids that constitute algal TGs are typically quite different from those of the polar lipids, EPA and other membrane-associated fatty acids will be diluted. Cohen et al. [51, 72] showed that cultivation of *P. cruentum* under nonoptimal conditions resulted in a decreased EPA content and a concomitant increase in AA and 18:2 in TGs. Similarly, Iwamoto and Sato [73] reported that increased NaCl concentration (from 0 to 3%) resulted in an elevation in the lipid content of the freshwater eustigmatophyte *M. subterraneus* from 34 to 46% (of dry weight) with a concurrent decrease in EPA from 34 to 15% (of total fatty acids). In the marine eustigmatophyte *Nannochloropsis,* the effect of salinity is rather different. At suboptimal NaCl concentrations, both growth and EPA production were predictably reduced [41]. However, increasing the salinity beyond the optimal level resulted in an increase in EPA yield despite a slight reduction in biomass.

Changes in lipid composition can also be observed as cultures age [74]. The highest levels of TGs and consequently the lowest levels of EPA are typically found at the stationary phase [75, 76]. The marine diatom *P. tricornutum* is however an exception, as the EPA level in this organism increased significantly as the cultures aged [52].

In another approach, Shimizu et al. [30] showed that supplementing the medium of EPA-producing fungi with linseed oil (rich in LNA) caused a doubling of the EPA levels of the fungi. Similarly, addition of 18:1, 18:2 or 18:3 to the cultivation medium of *Euglena gracilis* enhanced the production of PUFAs, especially AA and EPA [77]. However, the latter appears to be the only algal example of this kind. Boswell et al. [78] could not achieve any increase in the oil content of the diatom MK8908 following incubation with varios LNA-containing sources and similar results were observed in *P. cruentum* [Z. Cohen, unpubl. data].

Growth Temperature

An increase in the degree of membrane lipid fatty acid unsaturation in association with a decrease in growth temperature is an almost universal response and has been observed in cyanobacteria [79, 80], bacteria [81, 82], eukaryotic algae [41, 83, 84], yeasts [85], and fungi [33, 86, 87], as well as in higher plants. This response can be explained by the need to compensate for a decrease in membrane fluidity at lower temperatures [88]. Alternatively, Brown and Rose [85] postulated that reduced oxygen availability resulting from decreased solubility at elevated temperatures affects the oxygen-dependent fatty acid desaturases.

In microalgae, the mechanisms involved are uncertain, and the effects of temperature cannot be generalized since very few species have so far been studied and the results are not consistent between species. A decrease in EPA and the concomitant accumulation of shorter chain and more saturated apparent precursors in *Nannochloropsis* sp. at elevated temperatures led Seto et al. [41] to suggest that the elongation and desaturation enzymes involved in PUFA biosynthesis are thermolabile. A similar phenomenon was also observed in *P. tricornutum* [55]. In contrast, Kyle et al. [64] observed that while in diatoms the level of fatty acids

Table 2. Variations in growth rate, cellular fatty acid content and EPA productivity in *Nannochloropsis* sp. grown at different temperatures

Parameter	Temperature		
	18°C	25°C	32°C
Growth rate, doublings·day^{-1}	0.49	0.81	0.56
Total fatty acids, pg·cell^{-1}	1.86	0.89	0.64
EPA mol%	18.40	24.40	18.20
EPA content, pg·cell^{-1}	0.35	0.22	0.12
EPA productivity, mg·l^{-1}·day^{-1}	3.43	3.46	1.34

Cultures were grown batch wise at an average irradiance level of 150 µmol m^{-2}·s^{-1} PPFD [from Sukenik et al., 89].

containing up to two double bonds appeared to be modulated, as predicted, by the ambient temperature, that of long-chain PUFAs such as EPA was not.

From a practical point of view, fatty acid composition, total lipid content, and growth rate should be considered simultaneously in order to evaluate the potential for temperature modulation of fatty acids. In algal strains whose EPA content peaks at suboptimal growth temperatures, the EPA production rate (mg EPA d^{-1}) is severely reduced by increased temperature. Indeed, Sukenik et al. [89] have shown that the cellular content of total fatty acids and EPA in *Nannochloropsis* decreased as growth temperature increased (table 2) [89]. The relative amount of EPA had a maximal value at a temperature slightly below the optimum for growth. At supraoptimal temperatures, EPA productivity declined, most likely due to a low synthesis of fatty acids and a reduction in growth rate (table 2) [89]. In *P. cruentum,* the highest EPA proportion occurred at the optimal growth temperature (25°C), however, maximal fatty acid content was obtained at 30°C [51].

Light

In many algae, high light intensity has been reported to elevate the levels of PUFAs [51, 90, 91], which may result from an increase in oxygen availability and hence an increase in oxygen-requiring lipid desaturation rates [63].

In *P. cruentum* it was shown that an increase in light intensity affected an increase in the ω3:ω6 fatty acid ratio [51, 92]. However, several recent reports have demonstrated that in other algal species, high light intensity may have an opposite effect. This group includes the diatoms *Cyclotella, Nitzchia closterium*

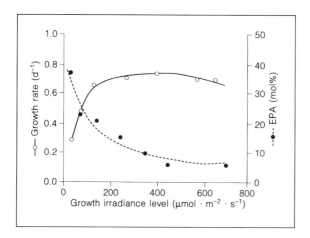

Fig. 3. Effect of irradiance level on cellular growth rate and EPA content in *Nanno-chloropsis* sp. grown in batch cultures. Cellular growth rate was monitored during the exponential growth phase, and EPA content (mol% of total fatty acids) was determined at the beginning of the stationary phase. Reproduced from Sukenik et al. [89].

[93, 94], *P. tricornutum* [73], *Chaetoceros calcitrans* and *Thalassiosira pseudonana* [95], and the prymnesiophyte, *Isochrysis galbana* [95]. Similar results were obtained by Sukenik [96] who reported that at growth-limiting light conditions, *Nannochloropsis* sp. contained high levels of EPA (fig. 3) [96] and preferentially synthesized GL. However, as growth became light saturated, 16:0 and 16:1 became predominant and more of neutral lipids, mainly TGs, were synthesized, indicating that GL synthesis rate and EPA content can be inversely related to irradiance level. In this case, the reduction in the proportion of EPA was somewhat compensated for by an increase in total fatty acid content. Furthermore, the EPA productivity of high-light-grown cultures was more than 40% higher than productivity under low light as a result of the differences in growth rate (table 3) [96].

Light-dark cycles may also have an effect on fatty acid composition. Sicko-Goad et al. [97] showed that in the diatom *Cyclotella meneghiniana,* the levels of EPA and other PUFAs were lowest in the early part of the light period and highest in the dark. They suggested that EPA accumulation precedes cell division which takes place at the end of the dark period or at the beginning of the light period.

Very few studies have reported the effect of light on DHA levels. In *I. galbana,* the relative distribution of DHA increased as irradiance level was increased from 30 to 300 µmol m^{-2}·s^{-1} PPFD, but slightly decreased at higher light intensities, while total lipid content decreased linearly [98].

Table 3. Effect of growth irradiance level on growth rate, cellular content of EPA and EPA productivity in *Nannochloropsis* sp.

Parameter	Growth irradiance level	
	LL	HL
Growth rate, day^{-1}	0.40	0.72
EPA, mol%	0.464	0.382
EPA content, pg·cell^{-1}	0.35	0.22
EPA productivity, mg·l^{-1}·day^{-1}	3.71	5.50

Steady-state cultures were grown in turbidostats at growth irradiances of 35 (LL) and 290 (HL) μmol·m^{-2}·s^{-1} PPFD and 25°C. EPA productivity was calculated according to equation 1 assuming a biomass concentration of 2 × 10^7 cells·ml^{-1} [from Sukenik, 96].

Nitrogen

The availability of nitrogen in the ambient is known to have an immense effect on lipid composition. Imposing nitrogen limitation when light is in excess results in cessation of growth and sometimes in an increase in the production of nonnitrogenous reserve materials such as TGs [99]. One study reported a 130–320% increase in oil content under nitrogen-deficient conditions in fifteen Chlorophyceae strains [100]. In most cases, this is accompanied by a decrease in the overall EPA content [72], since TGs are generally poor in EPA. Nitrogen deprivation in *P. cruentum* resulted, as predicted, in an increase in the proportion of TGs with a concomitant decrease in polar lipids; the level of EPA decreased in all lipids yet the proportion of AA increased from 20 to 31% in neutral lipids (NL) and from 46 to 61% in PL [51]. In the diatom MK8620, the EPA proportion in lipids decreased during nitrogen depletion while its content in the biomass increased due to the significant percentage of EPA in TGs in this organism [101]. However, in other cases, nitrogen limitation may provide a treatment to enhance PUFA content since several studies have reported significant proportions of various PUFA including EPA in the TG fraction of some algal species. Thus, the proportion of EPA in *Navicula saprophilla* exhibited little change following nitrogen deprivation while the total cellular lipid content nearly doubled resulting in a substantial increase in EPA content [64].

Silicon

In diatoms, which require silicon for their cell wall, a silicon deficiency was shown to induce lipid accumulation [102, 103] and at a much faster rate than under nitrogen deprivation. Roessler [104] demonstrated that in *Cyclotella cryptica* the percentage of newly assimilated carbon partitioned into lipids doubled within 4 h after the onset of silicon deficiency. Concomitantly, the activity of acetyl-CoA carboxylase doubled, and could be blocked by adding protein synthesis inhibitors [105]. These results suggest that an increase in the level of this enzyme, which may catalyze a rate-limiting step in fatty acid biosynthesis, can be induced by silicon deficiency, thus contributing to a higher capacity for lipid synthesis.

Distribution of Eicosapentaenoic Acid and Docosahexaenoic Acid in Microalgae

Long-chain ω3 PUFAs seem to be largely restricted to marine species; however, there have been a few reports of EPA occurrence in freshwater algae [73, 94, 106]. EPA and DHA are also absent from cyanobacteria and green algae (Chlorophyceae).

Members of the Bacillariophyceae (diatoms) include both freshwater and marine organisms; however, the only freshwater species found so far to contain EPA is *Phaeodactylum tricornutum*. EPA is quite ubiquitous in marine diatoms, and with very few exceptions it is also the major PUFA. However, a comparative study by Behrens et al. [107] showed that some of the species noted for a relatively high percentage of EPA in their lipids had a low overall oil content, while others produced large amounts of oil but had a low overall percentage of EPA (e.g. *Cylindrotheca fusiformis, Navicula pelliculosa* and the proprietary strain MK8620.

Some strains of the Eustigmatophyceae are among the leading algae with respect to their EPA content. The freshwater species *M. subterraneus* [73] and the unicellular marine alga *N. oculata* [41] contain 39 and 45%, respectively, of their total fatty acids as EPA.

Red algae (Rhodophyceae) are mostly macroalgae, and among the microalgae of this family, perhaps only *Porphyridium* and *Rhodella* have been studied with respect to fatty acids. The former is one of the few algae so far identified whose EPA proportion exceeds 40% of total fatty acids.

The PUFAs of Dinophyceae (dinoflagellates) include only 18:4, EPA and DHA. The unique characteristic of these algae is the relatively high DHA content which can range from 12 to 30% of total fatty acids (table 1) [45, 71, 108].

ω3 Fatty Acid Production by Phototrophic Algae

One objective of this review is to summarize the findings of studies designed to identify algal species which are the best potential sources of EPA under phototrophic conditions. In this section, some of those which have been investigated in most detail are discussed, with the reservation that the productivity of the EPA-producing strains was not always evaluated under culture conditions optimal for EPA synthesis.

Phaeodactylum tricornutum

Yongmanitchai and Ward [106] evaluated 17 algal species as potential sources for EPA. Growth rates and EPA contents were measured and EPA productivities were estimated. They concluded that the diatom *P. tricornutum* UTEX 640, formerly classified as *Nitzschia closterium* [109], was the most promising EPA producer. The proportion of EPA in this strain reached 31% of total fatty acids and its content was the highest among those strains tested, reaching 5.3% of dry weight. This strain also displayed the fastest growth rate, contributing to an EPA productivity rate of 19 mg $\cdot$ l$^{-1}\cdot$ d^{-1} which was the highest level found. In a later study, however, the maximum EPA content of this strain was reported to be only 3.3% [94].

Although *P. tricornutum* was originally reported as a marine diatom it requires neither NaCl nor silicates for growth or EPA production. The EPA level was found to be inversely proportional to cultivation temperature and light intensity [73]. EPA yields of up to 133 mg/l of culture were observed under optimized conditions [94]. The fatty acid composition of this strain is favorable as very low levels of other PUFAs are present. Furthermore, the comparatively large cell size facilitates rapid cell harvesting procedures.

EPA production in an outdoor culture of *P. tricornutum* was recently reported [110]. The EPA content of the biomass cultivated in a semicontinuous mode reached 3.9%. The biomass output rate was however rather low at 4.0 g$\cdot$ m$^{-2}\cdot$d^{-1}, resulting in an EPA yield of 0.15 g$\cdot$m$^{-2}\cdot$d^{-1}. In a batch mode, the EPA content decreased to 0.7%, and biomass production rate was halved.

Nannochloropsis oculata

N. oculata (Eustigmatophyceae), previously identified as marine *Chlorella,* is a unicellular alga which has been utilized as a biological feed in the aquaculture industry in Japan due to its exceptional nutritional characteristics [111]. The GL fraction of the cells contains over 70% EPA, indicating that the major molecular species is 20:5/20:5 [112]. Due to the absence of other PUFAs, the EPA from *N. oculata* can be easily purified to over 95% [113]. However, studies designed to optimize cultural conditions for EPA productivity have produced controversial results [41, 96]. As discussed above, Sukenik [96] has shown that the GL synthesis rate in *Nannochloropsis* sp., as well as the proportion of EPA can be inversely related to the irradiance level. Yet, the reverse relationship has been reported for batch cultures grown under relatively low light conditions [41, 113].

In synchronized cultures, cell division took place during the dark period, and fatty acid synthesis occurred in the light period [112]. The utilization of the more saturated TGs for maintenance in the dark led to an increase in the relative amount of EPA at the end of a diurnal cycle [96, 112].

Nannochloropsis is generally slow growing and is not oleogenic; however, Seto [113] reported an improved strain that was obtained by protoplast fusion of a slow-growing, EPA-containing strain with a faster growing freshwater *Chlorella*. The EPA level of brine shrimp was shown to rapidly elevate after feeding on this algal strain, which provided an excellent nutritional supplement to the cultured shrimp larvae. Dietary supplementation of spontaneously hypertensive rats with intact *Nannochloropsis* cells resulted in a significant reduction of the salt-induced rise in blood pressure. Furthermore, *Nannochloropsis*-fed zooplankton provided a highly nutritious live feed which significantly increased the survival of the hatchings of many cultured fish.

Under optimal growth conditions, cell densities of approximately $3 \cdot 10^7$ cells·ml^{-1} were obtained by Seto et al. [112, 114] in an open culture on a commercial scale. Even higher cell densities with elevated growth rates were maintained bacteria free in closed cultivation. Recently, Sukenik [pers. commun.] demonstrated the feasibility of large-scale outdoor cultivation of *Nannochloropsis* sp. in an overall production area of 1,700 m^2.

Monodus subterraneus

Several studies have indicated the potential value of *M. subterraneus* (Eustigmatophyceae) as a source of EPA. This alga belongs to an apparently very small group of freshwater species capable of EPA production. It contains EPA as the only major PUFA, other predominant fatty acids being 16:0 and 16:1 (table 1). Iwamoto and Sato [73] concluded that while total cellular lipid content increased with increasing temperature and light intensity, the proportion of EPA decreased. The alga grew well in a mixotrophic culture with acetate, but the EPA content was substantially lower. Cohen [unpubl. data] has shown that the maximum EPA production rate would be obtained at 25 °C under relatively high light intensity. Both of these investigations demonstrated a sharp decrease in EPA proportion of algal biomass under nitrogen starvation. In contrast, a Japanese patent [115] claimed that cultivation under high CO_2 followed by resuspension and holding in nitrogen-free medium elevated EPA in *M. subterraneus*.

Table 4. Effect of light intensity, cell concentration and temperature on the fatty acid composition of *P. cruentum*

Temp. °C	Light int.[a]	Cell conc.[b]	Growth rate[c]	Fatty acid composition, wt%										
				16:0	16:1	18:0	18:1	18:2	18:3	20:2	20:3	20:4	20:5	R
20	170	2	0.65	35.0	3.7	0.2	3.1	8.3	0.2	0.1	0.2	19.3	30.0	0.6
	170	28	stationary	32.7	1.8	0.5	1.0	8.3	0.3	1.2	1.2	28.8	24.1	1.2
25	170	1.5	1.2	35.0	2.8	0.6	2.0	6.1	0.5	0.3	0.9	15.9	35.8	0.4
	170	12	0.77	36.8	2.3	0.5	0.7	5.8	0.4	0.6	0.6	18.1	34.1	0.5
	170	30	stationary	33.3	1.4	0.3	1.0	9.0	0.2	0.7	1.7	27.3	25.1	1.1
30	170	1.5	0.98	38.9	2.9	1.1	1.5	11.0	1.0	0.5	0.8	18.6	23.7	0.8
	170	8	0.73	36.3	1.2	1.2	2.7	17.2	1.0	0.6	2.3	28.1	9.4	3.0
	30	3	0.30	33.9	0.1	0.1	2.3	12.0	0.1	0.2	1.5	42.1	7.8	5.4
	170	17	stationary	31.0	0.5	1.5	7.7	18.5	0.3	1.6	3.3	32.7	2.9	11.0

R = AA:EPA [from Cohen et al., 51].
[a] $\mu mol \cdot m^{-2} \cdot s^{-1}$ PPFD.
[b] mg chlorophyll$\cdot l^{-1}$.
[c] doublings$\cdot d^{-1}$.

Porphyridium cruentum

We have concentrated our research on the marine microalga *P. cruentum* as a source for both EPA and AA, taking advantage of desert environments where solar irradiation is abundant throughout the year and brackish water is plentiful. Moreover, cultivation on high-salinity medium may provide a unique ecological niche, thereby aiding the maintenance of monoalgal culture. Harvesting of this alga is less problematic than for some species as it has a tendency for autoflocculation, especially at low pH [116]. When considering the large-scale production of *Porphyridium* the possibility of obtaining other products from the biomass should also be emphasized. The presence of other products such as sulfated polysaccharides which can be used for enhanced oil recovery from oil wells [117], and phycoerythrin [118], a red proteinaceous pigment, should further encourage the outdoor cultivation of *Porphyridium*.

Cohen et al. [51, 72] showed that under conditions which allow the fastest exponential growth rate of *P. cruentum,* the dominant PUFA is EPA. In contrast, under growth-limiting conditions, AA and 18:2 become dominant, the magnitude of the effect being maximal at the stationary phase (table 4) [51]. These observations may explain some of the earlier studies reporting high levels of AA and little, if any, EPA in *Porphyridium* [55, 83, 119–121].

Table 5. Fatty acid composition of *P. cruentum* strains grown at 25°C

Strain	Growth phase	μ d⁻¹	Fatty acid composition, wt% of total fatty acids										Fatty acid content % of AFDW			
			16:0	16:1	18:0	18:1	18:2	18:3	20:2	20:3	20:4	20:5	TFA	AA	EPA	R
1380-1a	E	1.10	30.8	3.4	0.2	0.5	5.9	0.1	0.6	1.0	22.2	35.3	5.3	0.8	1.9	0.63
	S		30.8	2.1	0.6	1.1	9.6	–	1.4	1.5	33.5	19.3	5.8	1.9	1.1	1.7
1380-1b	E	1.13	30.3	3.0	0.3	0.4	5.2	0.7	0.8	0.5	15.2	43.5	5.6	0.9	2.1	0.43
	S		32.4	2.3	0.5	1.2	7.6	0.3	0.8	2.0	29.2	23.2	5.4	1.6	1.2	1.3
1380-1c	E	1.13	32.0	3.2	0.4	0.4	5.7	0.7	0.5	0.5	16.9	39.5	5.4	0.9	2.1	0.43
	S		33.4	2.3	0.8	1.2	7.6	0.2	0.7	2.2	23.8	27.8	5.4	1.3	1.5	0.86
1380-1d	E	1.15	30.6	3.5	0.1	0.3	4.9	0.6	0.6	0.4	14.7	44.1	5.5	0.8	2.4	0.33
	S		32.1	2.1	0.5	1.6	7.9	0.2	1.0	2.5	24.7	27.4	5.1	1.2	1.4	0.90
1380-1e	E	1.39	29.3	2.9	0.1	0.3	6.9	1.2	0.7	0.7	18.4	39.3	5.3	1.0	2.1	0.47
	S		29.7	1.8	0.7	2.3	11.4	0.6	0.7	2.9	33.5	16.3	5.7	1.9	0.9	2.1
1380-1f	E	1.10	29.7	2.9	0.1	0.7	6.2	1.0	0.7	0.6	17.8	40.1	5.2	0.9	2.1	0.44
	S		31.6	2.0	0.5	1.4	9.2	0.4	0.7	2.3	33.4	18.4	5.1	1.7	0.9	1.8
113.80	E	0.98	28.6	2.9	0.2	0.2	6.4	0.6	0.5	0.6	17.9	41.6	4.4	0.8	1.9	0.43
	S		33.3	1.4	0.4	1.0	9.0	0.3	0.7	1.7	27.3	25.1	5.1	1.4	1.3	1.10

TFA = wt% AFDW; R = AA:EPA; E = exponential phase, 4–5 mg chlorophyll·l⁻¹; S = stationary phase, 27–30 mg chlorophyll·l⁻¹ [from Cohen, 43].

As discussed previously, adverse conditions can either directly, or via an inhibition of growth, reduce production of EPA and GL in *Porphyridium.* Studies with outdoor cultivation have indicated that the fatty acid composition becomes primarily dependent on cell concentration [51]. This finding is of practical value since cell concentration under outdoor cultivation is more easily controlled than other parameters such as temperature. It seems feasible that either an AA or an EPA-rich biomass can be obtained by a close control of the cell concentration. Maintaining *Porphyridium* at low cell concentrations resulted in an enrichment in EPA [51]. In contrast, a higher content of AA was obtained when cell concentrations were kept high during the summer. The EPA proportion reached a maximum value (50% of fatty acids), corresponding to a content of 2.5% of ash-free dry weight (AFDW) in winter, while the concentration of AA peaked in the summer, reaching 30% of fatty acids and 1.4% of AFDW. However, the EPA output rate was primarily dependent on the biomass production rate and reached 0.28 g·m⁻²·d⁻¹ in the summer and 0.12 g·m⁻²·d⁻¹ in the winter. While cultivation temperature has been studied to some extent [51], the effects of such parameters as day length and light periodicity on PUFA composition are not yet clearly understood.

Studies designed to rate different strains of *P. cruentum* in terms of their potential to produce EPA should primarily consider three factors: (1) the growth rate under optimal conditions; (2) the EPA content (% of AFDW), and (3) the ratio of AA to EPA (R value) at the exponential and the stationary phases. The growth rate and EPA content determines the EPA production rate. The R value in the exponential phase reflects the extent of AA 'contamination', and consequently the degree of difficulty of EPA purification. The lower the R value, the easier the separation from AA. The difference between the corresponding R values for the exponential and stationary phases in the various strains can be used as an indicator of the expected variability in EPA contents under outdoor conditions, as a result of light and cell concentration changes. On this basis, we rated *P. cruentum* strain 1380–1d as the most promising from the standpoint of EPA productivity and purification [43]. This strain contained the highest level of EPA (2.4% of AFDW) and a low R value (0.33) in the exponential phase, which increased to only 0.90 in the stationary phase (table 5) [43].

Porphyridium produces AA in high quantities at 30°C under light limitation (low light or high cell concentration) or nitrogen starvation. At this temperature, strain 1380-1a was found to have the highest AA content in the exponential phase (2.0% of AFDW), while strains 1380-1e and 113.80 both reached an AA content of 2.5% of AFDW in the stationary phase.

Isochrysis galbana

The prymnesiophyte *I. galbana* is one of the few algae with a relatively high content of DHA. This organism is also one of the most commonly utilized phytoplankton species in mariculture systems, partly due to its DHA content [98] which in this case is primarily located in PLs [96]. The lipid content of *I. galbana* decreases with increasing light intensity, while the relative distribution of DHA increases, reaching a maximum of 17% at 300 μmol m$^{-2} \cdot$s^{-1} PPFD [98]. Under nitrogen limitation, lipid content tripled to 6 pg$\cdot$cell^{-1}, while the proportion of EPA decreased only slightly [96].

Selection of Herbicide-Resistant Clones

An approach for increasing the content of specific cell components is the use of inhibitors of selected steps in biosynthetic pathways. Generally, these compounds also inhibit growth. Some over-producers of the metabolite in question were found among higher plant lines selected for resistance to the growth inhibition [122–124]. Mutants of *Spirulina* showing elevated production of proline were obtained by selection in the presence of proline analogs [125]. Similar results were obtained in the cyanobacterium *Nostoc* [126] and in the alga *Nannochloris bacilaris* [127].

One compound of potential value in this approach is the substituted pyridazinone herbicide, SAN 9785, which, as discussed above, is an effective inhibitor of fatty acid desaturation [57]. It was shown that the effect on chloroplast LA desaturase is a direct inhibition of enzyme activity [128].

In the presence of 0.16 mM SAN 9785, the proportion of EPA in P. cruentum declined from 37% (of total fatty acids) to 31%, while that of AA was reduced only at higher concentrations (0.4 mM) [59]. The effect of SAN 9785 on fatty acid composition increased with herbicide concentration up to 0.4 mM. Predictably, the herbicide also inhibited cell growth with a maximal effect at 0.8 mM. The growth inhibition (via impairment of photosynthetic efficiency) probably results from the effect on chloroplast membrane composition, since PUFAs in chloroplast GLs are considered as structural components of the photosynthetic apparatus. Although we cannot rule out the possibility that growth may be adversely affected by a reduction in carbon assimilation resulting from a direct inhibition of photosynthesis by the herbicide.

We have hypothesized that one means to achieve resistance to inhibitors of lipid biosynthesis would be to elicit fatty acid overproduction. Obtaining resistance to the growth inhibition caused by such compounds could provide a basis for selection of algal lines capable of EPA overproduction. The growth inhibition rendered by SAN 9785 and the specific inhibitory effects it exerted on the production of EPA led us to utilize it for the selection of resistant *Porphyridium* cell lines some of which were EPA overproducers.

Resistance to SAN 9785 was built up by continuous cultivation in the presence of gradually increasing concentrations of the herbicide [59]. Cultures resistant to 0.16 mM SAN 9785 had, at this concentration, a much higher growth rate than the wild type in the same conditions and one almost as fast as the wild type cultivated in inhibitor-free medium. At a concentration of 0.4 mM, the growth of the resistant culture resembled that of the wild type in 0.16 mM SAN 9785, while the wild type collapsed after a few days. When transferred back to inhibitor-free medium, the growth rates of the resistant lines returned to control levels. The resistance was maintained even after 50 generations in an inhibitor-free medium, indicating an apparent change of a genetic nature.

The level of EPA in the presence of SAN 9785 reached 30.7% in resistant cultures while it was 26.5% in the wild type *Porphyridium* strain. A more focused view on the effect of SAN 9785 was obtained by comparing the fatty acid composition of the GL fraction which contains most of the EPA: the inhibitor reduced the proportion of EPA in the wild type from 50.0 to 29.1%, but only to 45.4% in the resistant culture (table 6) [129].

The resistance obtained could be the result of decreased uptake or enhanced metabolism of the inhibitor, or reduced affinity of the target enzyme, but could also emerge as a result of genetic changes leading to overproduction of the target

Table 6. Fatty acid composition of the GL fraction of wild type (1380-1d) and SAN-9785-resistant cultures of *P. cruentum*

Culture	Medium	Fatty acid composition, wt% of total fatty acids										
		16:0	16:1	16:3	18:0	18:1	18:2	18:3	20:2	20:3	20:4	20:5
WT	–	36.4	1.2	n.d.	0.5	1.0	4.3	n.d.	0.6	n.d.	5.9	50.0
WT	+	36.4	8.2	0.5	3.3	5.7	4.1	0.4	1.3	0.2	9.2	29.1
SRP[a]	+	36.0	2.0	0.2	1.2	2.0	7.4	0.3	1.2	0.2	3.8	45.4

– = Inhibitor free medium; + = medium containing 0.4 mM SAN 9785; n.d. = not determined [from Cohen et al., 129].

[a] A culture of *P. cruentum* 1380-1d resistant to 0.4 mM SAN 9785.

fatty acid. It appears that the latter was at least a contributing factor since elevated EPA levels were sustained for many generations after resistant cultures were shifted to herbicide-free medium. Following agar plating, 6 colonies were selected, 3 of which displayed even higher proportions of EPA and higher fatty acid contents, resulting in an overall 29% enhancement of the EPA content. These data support the proposed employment of inhibitors of fatty acid desaturation as means for obtaining fatty acid overproduction. To the best of our knowledge, this is the only report of fatty acid overproduction in either higher or lower plants induced by perturbation of fatty acid metabolism.

Further exploitation of this approach may be hampered by the low specificity of the herbicide. Thus, more specific inhibitors such as transition stage analogs are being sought. We anticipate that further selection of EPA-overproducing strains in conjunction with molecular biology techniques will allow manipulation of the regulation of relevant genes, thus, making algal mass production of PUFAs feasible. The establishment of herbicide resistance would also be advantageous for large-scale cultivation of *Porphyridium* since the herbicide may also be utilized in the maintenance of monoalgal cultures, which can be a major problem in continuous outdoor cultivation.

ω3 Fatty Acids Derived from Algal Triglycerides

Reported studies have largely concentrated on elevation of EPA in algal polar lipids. However, EPA containing TGs may be pharmaceutically preferable to the EPA ethyl ester which is the final form of EPA derived from polar lipids [130]. Generally, algal TGs are highly saturated and contain very little EPA. Although there are particular exceptions, and Kyle et al. [64, 131] have concentrated their efforts on cultivation and selection of oleaginous diatom strains which do contain EPA-enriched oil (TGs). A study was successful in the identification of diatom strain MK8620 which has the ability to convert EPA-containing polar lipids into TG, thereby enriching this lipid class in EPA [101].

Cultures of strain MK8620 were grown in a two-phase mode. In the first, nutrient-replete conditions allowed for exponential growth and biomass accumulation, while in the second phase, nitrogen was depleted resulting in enhanced TG accumulation in the form of large oil droplets. The overall effect was an increase in EPA-containing TGs [107]. Large-scale production was demonstrated in a 500-liter photobioreactor where biomass production rates of $0.6-0.8 \ g \cdot l^{-1} \cdot d^{-1}$ were obtained. In the oleaginous phase, a biomass of 2 kg, which yielded 700 g oil, was attained in a 2-week batching time [107].

Heterotrophic Eicosapentaenoic Acid and Docosahexaenoic Acid Production

In an attempt to circumvent problems associated with phototrophic cultivation of algae (including low cell concentrations and vulnerability to contamination), Boswell et al. [78] and Kyle et al. [131, 132] screened hundreds of microalgal strains for their ability to grow heterotrophically and produce an EPA-containing oil. Such species could be cultivated using conventional fermentation technology. One such isolate is strain MK8908, an apochlorotic diatom which grows well in a defined artificial seawater medium supplemented with glucose and silicate. In a fermentor, this microalga exhibited rapid growth rates and biomass productivity of about $15 \ g \cdot l^{-1} \cdot d^{-1}$. Cell concentrations as high as $40-50 \ g \cdot l^{-1}$ were obtained in 72 h [78, 131]. However, under conditions in which optimal oil production was obtained, EPA productivity was unacceptably low at only $0.12-0.14 \ g \cdot l^{-1} \cdot d^{-1}$ due to the low EPA content of the TGs. Current efforts are focussed on selection of mutants with a higher EPA content in their TGs (see previous section).

Heterotrophic DHA production by microalgae is also a future goal. Another proprietary diatom developed at the Martek laboratories is strain MK8805, which was also optimized to grow under heterotrophic conditions [133]. The polar lipids of this alga are highly enriched in DHA, and unlike most species, DHA was also a major component of TG. The oil content of this algal strain can amount to 35% of the biomass, making it extremely promising being rich in both oil and DHA. Moreover, since DHA is essentially the only PUFA, purification of this fatty acid should be neither difficult nor expensive [133].

Table 7. Purification of EPA methyl ester from *P. cruentum*

	Fatty acid composition, wt% of total fatty acids									
	16:0	16:1	18:0	18:1	18:2	18:3	20:3	20:4	20:5	R
Total lipids	30.6	5.3	0.3	0.9	4.8	0.8	2.0	14.6	40.7	0.36
FAME[a]	34.8	0.4	0.8	1.3	9.6	0.1	0.2	5.4	47.5	0.11
FAME after urea adduction	3.1	0.3	–	0.6	7.3	0.1	0.2	6.6	81.9	0.08
FAME after chromatography	0.3	–	–	–	0.9	–	0.2	1.4	97.3	0.01

FAME = Fatty acid methyl ester [from Cohen and Cohen, 134].

Eicosapentaenoic Acid Purification

At present, algal-derived EPA cannot compete economically with fish oil as a dietary supplement. However, once EPA can be introduced as a pharmaceutical commodity, the increased requirements will result in preference of sources from which purification would be the least costly. Algal oil may then be competitive, and particularly if the levels of EPA can be further increased as discussed above. The unique composition of algal oils offers distinct advantages for large-scale and efficient purification of EPA and other PUFAs.

Typically, algae contain EPA mainly in polar lipids and primarily in their GL fraction. Separation of this fraction from the lipid mixture is much easier than separation of EPA from other similar fatty acids, constituting a first purification step. This concept was tested on a *P. cruentum* biomass grown under temperature and light regimes most conducive to maximization of the EPA content (25°C, high light intensity and low cell concentration) [43]. Under these conditions, EPA constituted 40% of total fatty acids and AA 15%. Moreover, about 90% of the EPA was restricted to the GLs while AA was primarily contained in the TG and PL fractions. After an initial recovery of the GL fraction on silica, followed by transmethylation, and urea adduction (to remove saturated and monounsaturated fatty acids), the EPA concentration was increased to 82% [134]. The same procedure was utilized for the preparation of an AA concentrate from *P. cruentum* strain 1380-1b. These preliminary steps should provide an outstanding starting material for further purification by large-scale reverse-phase chromatography as recently demonstrated [134] (table 7). Using the simple techniques of urea adduction, solvent fractionation, and column chromatography, Seto et al. [113] have been able to produce EPA at 95–98% purity from *Nannochloropsis.*

Conclusions: Economic Considerations and Future Directions

While there is now an increasing demand for EPA- and DHA-containing oils in both the pharmaceutical and food industries, large-scale production of algal-derived EPA, DHA, and other PUFAs will depend on the introduction of EPA as a drug, as well as recognition of algal oil as a source of concentrated, pesticide-free PUFAs. The concept of producing 'designer algal oils' with PUFA compositions aimed for selected biomedical and commercial applications was recently introduced by Kyle et al. [132] and Kyle [135], and has already resulted in the development of supplementations of infant formula which optimize PUFA requirements [132]. Such developments will require further algal strain selection, and cultivation under closed fermentation conditions in order to reduce the future cost of biomass production.

In this review we have described the inherent limitations in using current approaches (physiological, strain selection, etc.) for obtaining enhanced EPA content in algae. Future attempts to increase algal EPA and DHA production will most likely include a full comprehension of the biosynthetic pathways involved as a prelude to genetic manipulation. The necessary steps will include the selection of mutants such as:

(1) Mutants deficient in their ability to produce EPA, DHA or their precursors will make possible the identification of regulation points (genes and enzymes) which participate in EPA and DHA biosynthesis. This approach using mutants of *A. thaliana* has previously enabled the elucidation of biosynthetic pathways of linolenic acid in higher plants [49]. Similar fatty-acid-deficient mutants were obtained from the cyanobacterium *Synechocystis* [136]. Researchers at the Martek Company, USA [131, 132] have even forecasted artificial combination of several desirable algal traits into a single algal strain via genetic engineering, and perhaps a gene transfer from PUFA-rich algae to other hosts, such as oleaginous yeasts, or to higher plants for agronomic production of PUFA-containing oil.

(2) Mutants which overproduce EPA can be obtained by selection of mutants resistant to inhibitors known to impair the biosynthesis of EPA.
Otherwise, clones with an improved fatty acid composition can be screened for at random. Thus Kyle et al. [131, 132] screened over 350 clones of the diatom strain MK8908 to identify those in the upper 5–10% level for EPA production.

(3) Herbicide-resistant mutants could be maintained in monoalgal cultures utilizing the herbicide as the selective pressure to prevent contamination. Several mutants of this nature were obtained from the green microalgae *Chlamydomonas reinhardtii* [137, 138]. A mutant resistant to the herbicide sulfometuron methyl has recently been isolated [139]. Of further relevance, Shifrin et al. [140] have identified copper-tolerant clones of *Chlorella vulgaris* and shown that the tolerance is also associated with enhanced oil production.

Alternative methods for identification of subpopulations of lipid overproducers are also becoming avilable [64, 101]. In diatoms which contain significant proportions of EPA within their TG fraction, one can introduce a nontoxic fluorescent dye such as Nile Red which partitions into the natural lipid fraction of a cell. By cytofluorometry, it is possible to select cells with enhanced fluorescence resulting from elevated lipid levels. Indeed, when cells of *N. saprophilla* were stained with Nile Red and analyzed, a normal distribution pattern of fluorescence was observed within a single population. This approach was also used by Solomon and Plumbo [141] to genetically improve lipid production in the microalgal species *I. galbana.*

A recent study [142] estimated the potential US market for ω3 fatty acids at 1,575 tons per annum, primarily for cardiovascular diseases, cancer, and arthritis. This forecast assumed a daily consumption of 3 g of these compounds by 2% of patients. The current cost of production of algal biomass on an industrial scale has been estimated by Chaumont et al. [143] at up to US$ 4,000/ton. Extraction of oils from dried algae was calculated as US$ 0.87/kg, of which capital charges would contribute about 60% [144]. Thus the cost of EPA production from an algal biomass containing 5% EPA would be about US$ 80/kg. As discussed here, due to the unique nature of algal oils, EPA purification can be achieved using low-cost techniques. In comparison, the cost of purification of EPA from fish oil in a pilot plant by super critical fluid extraction was estimated to be US$ 220/kg, exclusive of the feed material costs [145]. As previously emphasized [63], the availability of valuable co-products in microalgae (such as the pigmented protein phycoerythrin) should further reduce the cost of PUFA production.

References

1   Dyerberg J, Bang HO, Stoffersen E, et al: Eicosapentaenoic acid and prevention of thrombosis and atherosclerosis. Lancet 1975;ii:117–119.
2   Dyerberg J: Linoleate-derived polyunsaturated fatty acids and prevention of atherosclerosis. Nutr Rev 1986;44:125–134.
3   Mehta J, Lopez LM, Wargovich T: Eicosapentaenoic acid: Its relevance in atherosclerosis and coronary artery disease. Am J Cardiol 1987;59:155–159.
4   Urakaze M, Hamazaki T, Soda Y, et al: Infusion of emulsified trieicosapentaenoyl-glycerol into rabbits – The effect on platelet aggregation, polymorphonuclear leukocyte adhesion, and fatty acid composition in plasma and platelet phospholipids. Thromb Res 1986;44:673–682.
5   Phillipson BE, Rothrock DW, Connor WE, et al: Reduction of plasma lipids, lipoproteins, and aproteins by dietary fish oils in patients with hypertriglyceridemia. N Engl J Med 1985;312:1210–1216.
6   Mortensen JZ, Schmidt EB, Neilson AH, et al: The effect of n–6 and n–3 polyunsaturated fatty acids on hemostasis, blood lipids and blood pressure. Thromb Haemost 1983;50:543–546.
7   Braden LM, Carroll KK: Dietary polyunsaturated fat in relation to mammary carcinogenesis in rats. Lipids 1986;21:285–288.
8   Reddy BS, Maruyama H: Effect of dietary fish oil on azoxymethane-induced colon carcinogenesis in male F344 rats. Cancer Res 1986;46:3367–3370.

9   Lands WEM: The fate of polyunsaturated fatty acids; in Simopoulos AP, Kifer RR, Martin RE (eds): Health Effects of Polyunsaturated Fatty Acids in Seafoods. London, Academic Press, 1986, pp 33–48.

10 Kremer JM, Bigauaoette J, Michalek AV, et al: Effects of manipulation of dietary fatty acids on clinical manifestations of rheumatoid arthritis. Lancet 1985;i:184–187.

11 Higgs GA: The role of eicosanoids in inflammation. Prog Lipid Res 1986;25:555–561.

12 Ziboh VA, Cohen KA, Ellis CN, et al: Effects of dietary supplementation of fish oil on neutrophil and epidermal fatty acids. Modulation of clinical source of psoriatic subjects. Arch Dermatol 1986; 122:1277–1282.

13 Robinson DR, Prickett JD, Makoul GT, et al: Dietary fish oil reduces progression of established renal disease in (NZB × NZW)F$_1$ mice and delays renal disease in BXSB and MRL/1 strains. Arthritis Rheum 1986;29:539–546.

14 Samuelsson B: Leukotrienes: Mediators of immediate hypersensitivity reactions and inflammation. Science 1983;220:568–575.

15 Kobayashi S, Hirai A, Terano T, et al: Reduction in blood viscosity by eicosapentaenoic acid. Lancet 1981;ii:197.

16 Fischer S, Weber PC: Prostaglandin I$_3$ is formed in vivo in man after dietary eicosapentaenoic acid. Nature 1984;307:165–168.

17 Knapp HR: Fish oil supplements. Good for the heart? Drug Ther 1987;Feb:53–67.

18 Dratz EA, Deese AJ: The role of docosahexaenoic acid (22:6ω3) in biological membranes: Examples from photoreceptor and model membrane bilayers; in Simopoulos AP, Kifer RR, Martin RE (eds): Health Effects of Polyunsaturated Fatty Acids in Seafoods. London, Academic Press, 1986, pp 319–351.

19 Salem N Jr, Kim HY, Yargey JA: Docosahexaenoic acid membrane function and metabolism; in Simopoulos AP, Kifer RR, Martin RE (eds): Health Effects of Polyunsaturated Fatty Acids in Seafoods. London, Academic Press, 1986, pp 49–60.

20 Carlson SE, Cooke RJ, Werkman SH, et al: Long-term docosahexaenoic acid (DHA) and eicosapentaenoic (EPA) supplementation of preterm infants: Effects on biochemistry, visual acuity, information processing and growth in infancy. INFORM 1990;1:306. R3.

21 Uauy R, Birch D, Birch E, et al: Are omega-3 fatty acids essential for eye and brain development in humans? INFORM 1990;1:306. R4.

22 Carlson SE, Peeples JM, Werkman JM, et al: Arachidonic acid (AA) in plasma and red blood cell (RBC) phospholipids (PL) during followup of pre-term infants: Occurrence, dietary determinants and functional relationships. 2nd International Conference on the Health Effects of Omega-3 Polyunsaturated Fatty Acids in Seafoods, Washington, March 1990.

23 Schweikhardt F: German Patent No. 3, 603,000 (1987).

24 Ackman RG: Fatty acid composition of fish oils; in Barlow SM, Stansby ME (eds): Nutritional Evaluation of Long-Chain Fatty Acids in Fish Oil. London, Academic Press, 1982, pp 25–88.

25 Bimbo AP: The emerging marine oil industry. J Am Oil Chem Soc 1987;64:706–715.

26 Yazawa K, Araki K, Okazaki N, et al: Production of eicosapentaenoic acid by marine bacteria. J Biochem 1988;103:5–7.

27 Sargent JR, Bell MV, Henderson RJ, et al: in Mellinger J (ed): Comp Physiology. Basel, Karger, 1990, vol 5, pp 11–23.

28 Klausner A: Algaculture: Food for thought. Biotechnology 1985;4:947–953.

29 Lie O: Influence of dietary fatty acids on the composition of neutral lipids and glycerolipids of erythrocytes and spleen in cod (Gadus morhua). INFORM 1991;2:310.

30 Shimizu S, Akimoto K, Kawashima H, et al: Microbial conversion of an oil containing alpha-linolenic acid to an oil containing eicosapentaenoic acid. J Am Oil Chem Soc 1989;66:342–347.

31 Shimizu S, Shinmen Y, Kawashima H, et al: Fungal mycelia as a novel source of eicosapentaenoic acid production at low temperature. Biochem Biophys Res Commun 1988;150:335–341.

32 Totani N, Oba K: The filamentous fungus Mortierella alpina, high in arachidonic acid. Appl Microbiol Biotechnol 1988;28:135–137.

33 Shimizu S, Kawashima H, Shinmen Y, et al: Production of eicosapentaenoic acid by Mortierella fungi. J Am Oil Chem Soc 1988;65:1455–1459.

34   Burlew JS: Current status of the large-scale culture of algae; in Burlew JS (ed): Algal Culture from Laboratory to Pilot Plant. Washington DC, Carnegie Institution of Washington, 1953, publ No 600, pp 3–23.

35   Wood JB: Fatty acids and saponifiable lipids; in Stewart WDP (ed): Algal Physiology and Biochemistry. Oxford, Blackwell 1974, pp 236–265.

36   Volkman JK, Jeffrey SW, Nichols PD, et al: Fatty acid and lipid composition of 10 species of microalgae used in mariculture. J Exp Marine Biol Ecol 1989;128:219–240.

37   Pohl P: Lipids and fatty acids of microalgae; in Zaborsky O (ed): CRC Handbook of Biosolar Resources. Boca Raton, CRC Press, 1982, vol 1/pt 1, pp 383–404.

38   DeMort CL, Lowry R, Tinsley I, Phinney HK: The biochemical analysis of some estuarine phytoplankton species. I. Fatty acid composition. J Phycol 1972;8:211–216.

39   Gillan FT, McFadden R, Wetherbee R, Johns RB: Sterols and fatty acids of an Antarctic Sea ice diatom, *Stauroneis amphioxys.* Phytochemistry 1981;20:1935–1937.

40   Cobelas MA, Lechado JZ: Lipids in microalgae. A review. I. Biochemistry. Grasas Aceites 1989;40:118–145.

41   Seto A, Wang HL, Hesseltine CW: Culture conditions affect eicosapentaenoic acid content of *Chlorella minutissima.* J Am Oil Chem Soc 1984;61:892–894.

42   Ackman RG, Tocher CS: Marine phytoplantkon fatty acids. J Fish Res 1968;25:1603–1620.

43   Cohen Z: The production potential of eicosapentaenoic and arachidonic acids by the red alga *Porphyridium cruentum.* J Am Oil Chem Soc 1990;67:916–920.

44   Pohl P, Zurheide F: Fat production in freshwater and marine algae; in Hoppe HA, Levring T, Tanaka Y (eds): Marine Algae in Pharmaceutical Science. Berlin, Walter de Gruyter, 1982, pp 65–80.

45   Joseph JD: Identification of 3, 6, 9, 12, 15–octadecapentaenoic acid in laboratory-cultured photosynthetic dinoflagellates. Lipids 1975;10395–10403.

46   Frentzen M: Biosynthesis and desaturation of the different diacylglycerol moieties in higher plants. J Plant Physiol 1986;124:193–209.

47   Heemskerk JWM, Wintermans JFGM: Role of the chloroplast in the leaf acyl lipid synthesis. Physiol Plant 1987;70:558–568.

48   Harwood JL: Fatty acid metabolism. Annu Rev Plant Physiol Plant Mol Biol 1988;39:101–138.

49   Browse J, Miquel M, Somerville C: Genetic approaches to understanding plant lipid metabolism; in Quinn PJ, Harwood JL (eds): Plant Lipid Biochemistry, Structure and Utilization. London, Portland Press, 1990, pp 431–438.

50   Araki S, Sakurai T, Oohusa T, et al: Content of arachidonic and eicosapentaenoic acids in polar lipid from *Gracilaria* (Gracilariales, Rhodophyta). Hydrobiologia 1990;204/205:513–519.

51   Cohen Z, Vonshak A, Richmond A: Effect of environmental conditions on fatty acid composition of the red alga *Porphyridium cruentum:* Correlation to growth rate. J Phycol 1988;24:328–332.

52   Arao T, Kawaguchi A, Yamada M: Positional distribution of fatty acids in lipids of the marine diatom *Phaeodactylum tricornutum.* Phytochemistry 1987;26:2573–2576.

53   Kates M, Volcani BE: Lipid components of diatoms. Biochim Biophys Acta 1966;11:264–278.

54   Arao T, Yamada M: Positional distribution of polyunsaturated fatty acids in galactolipids from some algae of Rhodophyta, Phaeophyta, Bacillariophyta and Chlorophyta; in Biacs PA, Gruiz K, Kremmer T (eds): Biological Role of Plant Lipids. New York, Plenum Publishing/Budapest, Academiai Kiado, 1989, pp 265–266.

55   Moreno VJ, De Mareno JEA, Brenner RR: Biosynthesis of unsaturated fatty acids in the diatom *Phaeodactylum tricornutum.* Lipids 1979;14:15–19.

56   Nichols BW, Appleby RS: The distribution and biosynthesis of arachidonic acid in algae. Phytochemistry 1969;8:1907–1915.

57   Hoppe HA: Fatty acid biosynthesis – A target site of herbicide action; in Boger P, Sandman G (eds): Target Sites of Herbicide Action. Boca Raton, CRC Press, 1989, pp 65–84.

58   Norman HA, St John JB: Differential effects of a substituted pyridazinone, BASF 13–338 on pathways of monogalactosyldiacylglycerol synthesis in Arabidopsis. Plant Physiol 1987;85:684–688.

59   Cohen Z, Didi S, Heimer YM: Overproduction of gamma-linolenic and eicosapentaenoic acids by algae. Plant Physiol 1992;98:569–572.

60 Norman HA, St John JB: Metabolism of unsaturated monogalactosyldiacylglycerol molecular species in *Arabidopsis thaliana* reveals different sites and substrates for linolenic acid synthesis. Plant Physiol 1986;81:731–736.

61 Henderson RJ, Hodgson P, Harwood JL: Differential effects of the substituted pyridazinone herbicide Sandoz 9785 on lipid composition and biosynthesis in non-photosynthetic marine microalgae. 1. Lipid composition and synthesis. J Exp Bot 1990;41:729–736.

62 Henderson RJ, MacKinlay EE: Effect of temperature on lipid composition of the marine cryptomonad *Chromonas salina.* Phytochemistry 1989;28:2943–2948.

63 Kyle DJ: Microbial omega-3-containing fats and oils for food use. Adv Appl Biotechnol 1991;12: 167–183.

64 Kyle DJ, Behrens PW, Bingham S, et al: Microalgae as a source of EPA-containing oils; in Applewhite TH (ed): Proceedings of the World Conference on Biotechnology for the Fats and Oils Industry. Am Oil Chem Soc, Champaign, 1989, pp 117–122.

65 Chen H, Bingham SE, Chantler V, et al: ^{14}C-Labeled fatty acids from microalgae. Dev Ind Microbiol 1990;31:257–264.

66 Kyle DJ, Glaued R: Microalgae as a source of EPA; in Ekstrand B (ed): New Aspects of Dietary Lipids. Benefits, Hazards and Uses. Göteborg, SAK, 1990, pp 161–170.

67 Gellerman JL, Schlenk H: Methyl directed desaturation of arachidonic acid to eicosapentaenoic acid in the fungus *Saprolegnia parasitica.* Biochim Biophys Acta 1979;573:23–30.

68 Hodgson PA, Henderson RJ, Sargent JR, et al: Patterns of variation in the lipid class and fatty acid composition of *Nannochloropsis oculata* (Eustigmatophyceae) during batch culture. 1. The growth cycle. J Appl Phycol 1991;3:169–181.

69 Sukenik A, Livne A: Variations in lipid and fatty acid content in relation to acetyl CoA carboxylase in the marine prymnesiophyte *Isochrysis galbana.* Plant Cell Physiol 1991;32:371–378.

70 Henderson RJ, MacKinley EE: Polyunsaturated fatty acid metabolism in the marine dinoflagellate *Crypthecodinium cohnii.* Phytochemistry 1991;30:1781–1787.

71 Henderson RJ, Leftley JW, Sargent JR: Lipid composition and biosynthesis in the marine dinoflagellate *Crypthecodinium cohnii.* Phytochemistry 1988;27:1679–1683.

72 Cohen Z, Vonshak A, Richmond A: The effect of environmental conditions on fatty acid composition of *Porphyridium cruentum;* in Stumpf PK, Mudd JB, Ness WD (eds): The Metabolism, Structure and Function of Plant Lipids. New York, Plenum Press, 1986, pp 641–644.

73 Iwamoto H, Sato S: Production of EPA by freshwater unicellular algae. J Am Oil Chem Soc 1986; 63:434.

74 Cohen Z: Products from microalgae; in Richmond A (ed): Handbook for Microalgal Mass Culture. Boca Raton, CRC Press, 1986, pp 421–454.

75 Iwamoto H, Yonekawa G, Asai T: Fat synthesis in unicellular algae. I. Culture conditions for fat accumulation in *Chlorella* cells. Bull Agric Chem Soc Jpn 1955;19:240–246.

76 Klyachko-Gurvich GL, Zhukova TS, Vladimirova MG, et al: Comparative characteristics of the growth and direction of biosynthesis of various strains of *Chlorella* under nitrogen deficiency conditions. III. Synthesis of fatty acids. Soviet Plant Physiol 1967;16:205–209.

77 Okumura M, Ii S, Fujii R, et al: Jpn Kokai, Tokyo Koho 1986;JP61:177, 990.

78 Boswell KDB, Glaued R, Prima B, et al: SCO prodcution by fermentative microalgae; in Kyle DJ, Ratledge C (eds): Industrial Applications of Single Cell Oils. Am Oil Chem Soc, Champaign, 1992, pp 274–286.

79 Holton RW, Blacker HH, Onure M: Effect of growth temperature on the fatty acid composition of a blue-green alga. Phytochemistry 1964;3:595–602.

80 Kleinschmidt MG, McMahon VA: Effect of growth temperature on the lipid composition of *Cyanidium caldarium.* II. Glycolipid and phospholipid components. Plant Physiol 1970;46:290–293.

81 Duran HH: Fatty acid composition of lipid extracts of a thermophilic bacillus species. J Bacteriol 1970;101:145–151.

82 Kates M, Hagen PO: Influence of temperature on fatty acid composition of psychrophilic and mesophilic *Serratia* species. Can J Biochem 1964;42:481–488.

83 Ahern TJ, Katoh JS, Sada E: Arachidonic acid production by the red alga *Porphyridium cruentum.* Biotechnol Bioeng 1983;25:1057–1070.

84 Patterson GW: Effect of culture temperature on fatty acid composition of *Chlorella sorokiniana.* Lipids 1970;5:598–600.

85 Brown CM, Rose AH: Fatty acid composition of *Candida utilis* as affected by growth temperature and dissolved oxygen tension. J Bacteriol 1969;99:371–376.

86 Hansson L, Dostalek M: Effect of culture conditions on mycelial growth and production of gamma-linolenic acid by the fungus *Mortierella ramanniana.* Appl Microbiol Biotechnol 1988;28:240–246.

87 Sumner JL, Morgan ED, Evans HC: The effect of growth temperature on the fatty acid composition of fungi in the order Mucorales. Can J Microbiol 1969;15:515–520.

88 Marr AG, Ingraham LJ: Effect of temperature on the composition of fatty acids in *E. coli.* J Bacteriol 1962;84:1260–1268.

89 Sukenik A, Carmeli Y, Berner Y: Regulation of fatty acid composition by growth irradiance level in the eustigmatophyte *Nannochloropsis* sp. J Phycol 1989;25:686–692.

90 Hitchcock C, Nichols BW: Plant Lipid Biochemistry. London, Academic Press, 1971.

91 Smith KL, Harwood JL: Lipid metabolism in *Fucus serratus* as modified by environmental factors. J Exp Bot 1984;35:1359–1368.

92 Siron R, Giusti G, Berland B: Changes in the fatty acid composition of *Phaeodactylum tricornutum* and *Dunaliella tertiolecta* during growth under phosphorus deficiency. Marine Ecol Prog Ser 1989; 55:95–100.

93 Orcutt DM, Patterson GW: Effect of light intensity upon lipid composition of *Nitzschia closterium (Cylindrotheca fusiformis).* Lipids 1974;9:1000–1003.

94 Yongmanitchai W, Ward OP: Growth of and omega-3 fatty acid production by *Phaeodactylum tricornutum* under different culture conditions. Appl Environ Microbiol 1991;57:419–425.

95 Harrison PJ, Thompson PA, Calderwood GS: Effects of nutrient and light limitation on the biochemical composition of phytoplankton. J Appl Phycol 1990;2:45–56.

96 Sukenik A: Ecophysiological considerations in the optimization of eicosapentaenoic acid production by *Nannochloropsis* sp. (Eustigmatophyceae). Bioresource Technol 1991;35:263–269.

97 Sicko-Goad L, Simmons MS, Lazinsky D, et al: Effect of light cycle on diatom fatty acid composition and quantitative morphology. J Phycol 1988;24:1–7.

98 Sukenik A, Wahnon R: Biochemical quality of marine unicellular algae with specific emphasis on lipid composition. I. *Isochrysis galbana.* Aquaculture 1991;97:61–72.

99 Mayzaud P, Chanut JP, Ackman RG: Seasonal changes of the biochemical composition of marine particulate matter with special reference to fatty acids and sterols. Marine Ecol Prog Ser 1989;56: 189–204.

100 Shifrin NS, Chisholm SW: Phytoplankton lipids: Interspecific differences and effects of nitrate, silicate, and light-dark cycles. J Phycol 1981;17:374–384.

101 Hoeksema SD, Behrens PW, Glaued R, et al: An EPA-containing oil from microalgae in culture; in Chandra RK (ed): Health Effects of Fish Oils. Newfoundland, ARTS Biomedical, 1989, pp 337–347.

102 Werner D: Die Kieselsäure im Stoffwechsel von *Cyclotella cryptica* Reimann, Lewin, und Guillard. Arch Mikrobiol 1966;55:278–308.

103 Coombs J, Darley WM, Holm-Hansen O, et al: Studies on the biochemistry and fine structure of silica shell formation in diatoms. Chemical composition of *Navicula pelliculosa* during silicon-starvation synchrony. Plant Physiol 1967;42:1601–1606.

104 Roessler PG: Changes in the activities of various lipid and carbohydrate biosynthetic enzymes in the diatom *Cyclotella cryptica* in response to silicon deficiency. Arch Biochem Biophys 1988;267:521–528.

105 Roessler PG: Effects of silicon deficiency on lipid composition and metabolism in the diatom *Cyclotella cryptica.* J Phycol 1988;24:394–400.

106 Yongmanitchai W, Ward OP: Screening of algae for potential alternative sources of eicosapentaenoic acid. Phytochemistry 1991;30:2963–2967.

107 Behrens PW, Hoeksema SD, Arnett KL, et al: Eicosapentaenoic acid from microalgae; in Demain AL, Somkuti GA, Hunter-Cevera JC, Rossmore HW (eds): Novel Microbial Products for Medicine and Agriculture. Amsterdam, Elsevier Science, 1989, pp 253–259.

108 Harrington GW, Beach DH, Dunham JE, et al: The polyunsaturated fatty acids of marine dinoflagellates. J Protozool 1970;17:213.
109 Lewin JC, Lewin RA, Philputt DE: Observations of *Phaeodactylum tricornutum.* J Gen Microbiol
1958;18:418–426.
110 Veloso V, Reis A, Gouveia L, et al: Lipid production by *Phaeodactylum tricornutum.* Bioresource
Technol 1991;38:115–119.
111 Watanabe T, Oowa F, Kitajima C, et al: Relationship between dietary value of brine shrimp *Artemia
salina* and their content of omega-3 highly unsaturated fatty acids. Bull Jpn Soc Sci Fish 1980;46:
35–41.
112 Sukenik A, Carmeli Y: Lipid synthesis and fatty acid composition in *Nannochloropsis* sp. (Eustigmatophyceae) grown in a light-dark cycle. J Phycol 1990;26:463–469.
113 Seto A, Kumasaka K, Hosaka M, et al: Production of eicosapentaenoic acid by a marine microalga
and its commercial utilization for aquaculture; in Kyle DJ, Ratledge C (eds): Single Cell Oil. Champaign, Am Oil Chem Soc, 1992, pp 219–234.
114 Seto A: Production of eicosapentaenoic acid by a marine microalga and its commercial utilization
for aquaculture. Proceedings of ISFJOCS World Congress, 1988, vol 2, p 1283.
115 Patent: Meiji Milk Products Co, Ltd, Japan: Eicosapentaenoic acid from algae. Jpn Kokai, Tokyo
Koho 1985;JP60:87, 798 (Chem Abstr 103:121762).
116 Vonshak A, Cohen Z, Richmond A: The feasibility of mass culture of *Porphyridium.* Biomass 1985;
8:13–25.
117 Thepenier C, Gudin C: Immobilization of *Porphyridium cruentum* in polyurethane foams for the
production of polysaccharide. Biomass 1985;7:225–240.
118 Glazer AN, Stryer L: Phycofluor probes. Trends Biochem Sci 1984;9:423–427.
119 Opute FI: Lipid and fatty-acid composition of diatoms. J Exp Bot 1973;25:823–835.
120 Kost HP, Senser M, Wanner G: Effect of nitrate and sulphate starvation on *Porphyridium cruentum*
cells. Z Pflanzenphysiol 1984;113S:231.
121 Lee YK, Tan HM, Low CS: Effect of salinity of medium on cellular fatty acid composition of marine
alga *Porphyridium cruentum* (Rhodophyceae). J Appl Phycol 1989;1:19–23.
122 Widholm JM: Selection and characterization of amino acid analog-resistance in plant cell culture.
Crop Sci 1977;17:597–600.
123 Maliga P: Isolation and characterization of mutants in plant cell culture. Int Rev Cytol
1980(suppl 11A):225–249.
124 Hibberd KA, Walter T, Green CE, et al: Selection and characterization of feed-back in sensitive
tissue culture of maize plants. Planta 1980;148:183–187.
125 Riccardi G, Sora S, Cifferi O: Production of amino acids by analog-resistant mutants of the cyano-
bacteriun *Spirulina platensis.* J Bacteriol 1981;147:1002–1007.
126 Kumar HD, Tripathi AK: Isolation of hydroxyproline-secreting pigment-defficient mutants of *Nostoc* sp. by metronosazole selection. MIRCEN J 1985;1:269–275.
127 Vanlerberghe GC, Brown LM: Proline overproduction in cells of the green alga *Nannochloris bacil-
laris* resistant to azetidine-2-carboxylic acid. Plant Cell Environ 1987;10:251–257.
128 Norman HA, Pillai P, St. John JB: In vitro desaturation of molecular species of monogalactosyldiacylglycerol and phosphatidylcholine by soybean chloroplast homogenates. Phytochemistry 1991;30:
2217–2222.
129 Cohen Z, Reungjitchachawali M, Siangdung W, et al: Herbicide resistant lines of microalgae:
Growth and fatty acid composition. Phytochemistry 1993;34:973–979.
130 Hamazaki T, Urakaze M, Makuta M, et al: Intake of different eicosapentaenoic acid-containing
lipids and fatty acid pattern of plasma lipids in rats. Lipids 1987;22:994–998.
131 Kyle DJ, Boswell KDB, Glaued RM, et al: Designer oils from microalgae as nutritional supplements;
in Bills DD, Kung SD, Kotula A, Watada A, Westhoff D, Quebedeaux B, Fader G, Goss J (eds):
Biotechnology and Nutrition. Proceedings of the Third International Symposium. Boston, Butter-
worth-Heinemann, 1991, pp 451–468.
132 Kyle DJ, Glaued R, Reeb S, et al: Production and use of omega-3 designer oils from microalgae; in
Developments in Industrial Microbiology Extended Abstract Series. Short Communications of the
1991 International Biotechnol Conference. Dubuque, Iowa, Brown, 1993, vol 2, pp 473–480.

133 Kyle DJ, Sicotte VJ, Reeb SE: Bioproduction of docosahexaenoic acid (DHA) by microalgae; in Kyle DJ, Ratledge C (eds): Single Cell Oil. Champaign, Am Oil Chem Soc, 1992, pp 287–320.

134 Cohen Z, Cohen S: Preparation of eicosapentaenoic acid (EPA) concentrate from *Porphyridium cruentum.* J Am Oil Chem Soc 1991;68:16–19.

135 Kyle DJ: Specialty oils from microalgae: New perspectives; in Rattray J (ed): Biotechnology of Plant Fats and Oils. Champaign, Am Oil Chem Soc, 1991, pp 130–143.

136 Wada H, Murata N: *Synechocystis* PCC6803 mutants defective in desaturation of fatty acids. Plant Cell Physiol 1989;30:971–978.

137 Galloway RE, Mets L: Atrazine, bromacil, and diuron resistance in *Chlamydomonas.* A single non-mendelian genetic locus controls the structure of the thylakoid binding site. Plant Physiol 1984;74: 469–474.

138 Erickson JM, Rahire M, Bennoun P, et al: Herbicide resistance in *Chlamydomonas reinhardtii* results from a mutation in the chloroplast gene for the 32-kilodalton protein of photosystem II. Proc Natl Acad Sci USA 1984;81:3617–3621.

139 Van-Moppes D, Barak Z, Chipman D, et al: An herbicide (sulfometuron methyl) resistant mutant in *Porphyridium* (Rhodophyta). J Phycol 1989;25:108–112.

140 Shifrin N; in Ratledge C, Dawson P, Rattray J (eds): Biotechnology for the Fats and Oils Industry. Champaign, Am Oil Chem Soc, 1984, pp 145–162.

141 Solomon JA, Plumbo AV: Improvement of microalgal strains for lipid production; in Johnson DA, Sprague S (eds): Solar Energy Research Institute Annual Report. Golden, 1987, pp 142–154.

142 Van der Wende LA: Omega-3 fatty acids in Biomarkets: 33 Market Forecasts for Key Product Areas. New Jersey, Technical Insights Inc, 1988, pp 199–203.

143 Chaumont D, Thepenier C, Gudin C, et al: Scaling up a tubular photoreactor for continuous culture of *Porphyridium cruentum* from laboratory to pilot plant (1981–1987); in Stadler T, Mollion J, Verdus MC, et al (eds): Algal Biotechnology. London, Elsevier Applied Science, 1988, pp 199–208.

144 Regan DR, Gartside G: Liquid fuels from microalgae in Australia. Melbourne, Commonwealth Scientific and Industrial Research Organization (CSIRO), 1983.

145 Latta S: Supercritical fluids attracting new interest. INFORM 1990;1:810–816.

Zvi Cohen, PhD, The Laboratory for Microalgal Biotechnology, Jacob Blaustein Institute for Desert Research, Ben-Gurion University of the Negev, Sede-Boker Campus, 84990 (Israel)

Simopoulos AP (ed): Plants in Human Nutrition.
World Rev Nutr Diet. Basel, Karger, 1995, vol 77, pp 32–46

....

Nutritional Value of the Alga Spirulina

J.C. Dillon[a], *Anh Phan Phuc*[b], *J.P. Dubacq*[c,1]

[a] Institut National Agronomique, Nutrition Humaine,
[b] CNRS, Laboratoire des biomembranes des cellules végétales, UA CNRS 1180 et
[c] Ecole Normale Supérieure, Paris, France

Contents

History of Spirulina 33
Production 33
Chemical Composition 34
    Composition and Nutritional Value of Proteins 35
    Composition and Nutritional Value of Lipids 38
    Vitamins 41
    Minerals 42
Spirulina as a Human Food 43
Conclusion 44
References 44

Spirulinas are cyanobacteria living in high salt alkaline water in subtropical and tropical areas. These algae are filamentous and measure 250 µm in length. Under the microscope, they appear as blue-green filaments due to the presence of both chlorophyll (green) and phycocyanin (blue) pigments in the cells. The prolific reproductive capacity of the cells and their proclivity to adhere in colonies makes spirulinas a large and easily gathered plant mass

There are 35 species of spirulina, proliferating in the wild in several alkaline lakes of the world: Lake Chad (Chad), Lake Texcoco (Mexico), and Lake Turkana (Kenya). Since the spiral shapes of the algae seem to metamorphose spontaneously depending on pH and nutrient conditions, it is possible that the different mor-

[1] The authors wish to thank Dr. Claude Sautier for providing many documents and Mr. Durand-Chastel for his moral support and constructive comments.

phologies (hence their names: *Spirulina platensis, Spirulina maxima* ...) are simply variations of a single species.

Spirulinas are potentially of considerable importance in human nutrition due to their overall nutritional qualities: high protein content (60–70% of dry weight), low fat, high vitamin and high γ-linolenic acid. Furthermore, due to the absence of cellulose cell walls, spirulina does not require chemical or physical processing steps in order to become digestible. Removal of moisture by simple sun-drying is sufficient as postharvesting treatment.

Today spirulina is cultivated in several countries with a total production of a few hundred tons for the health food market. In a few developing countries, spirulina is produced under local conditions based on simple techniques in rural areas.

History of Spirulina

The history of spirulina as a staple in the human diet is unique. There is evidence from the annals of the Spanish conquest of Mexico, early in the 16th century, that the Aztecs harvested mats of algal biomass from Lake Texcoco reminiscent of spirulina, from which they made bricks which were eaten just as cheese is eaten today in the Western world [1].

Likewise, the Kanembu tribe living along the shores of Lake Chad in Central Africa used dried spirulina as a food [2–5]. The alga, growing naturally in hot alkaline water, was harvested and sun-dried in the form of cakes that were used in the preparation of a dressing for local dishes.

Clément [6, 7], at the Institut Français des Pétroles, isolated a strain of *S. platensis* and studied the effect of various physical and chemical conditions on the growth of this alga. Several pilot units were built for studying algal production with combustion gases containing 10–13% of CO_2 as the carbon source [8]. A large-scale harvesting and processing plant was also developed [9].

An industrial plant at Sosa Texcoco, capable of producing 1 ton of dried *S. maxima* per day was later reported operational in 1972. The alga grows naturally in Lake Texcoco which is part of the valley of Mexico City located at 2,200 m above sea level. The present Mexico process includes filtration, fluidization, pasteurization, homogenization and drying [10, 11].

Production

The industrial production of spirulina requires sunlight, CO_2, mineral salts and water [12, 13]. The most favorable locations for algal production are found anywhere between latitudes 35°N and S. of the equator where solar flux is high, temperature variations are not excessive, rainfall moderate and source of CO_2 available. The utilization of combustion gases as a source of CO_2 in pilot-plant studies was successful [8]. However, chemicals for mineral salts usually constitute an important portion of the production cost for algae. Attempts to use effluents from secondary wastewater treatment plants as an economical source of nitrogen and phosphate salts for spirulinas have been reported [14, 15].

Much progress has been accomplished in the last 20 years in developing the biotechnology for algal mass culture [16–19]. Improvements have been made mainly in the management of outdoor cultures based on a better understanding of the biology of dense cultures grown on a large scale as well as of technological aspects, such as pond design, mixing systems (sea water, urea, biogas digester) and mode of harvesting [20–23].

In large-scale plants annual production does not exceed 30 ton/ha/year, representing an average of less than 10 g/m^2/day. The future of this industry is greatly dependent on the development of the consumer market in the Western world and should result in increased yield with a corresponding decrease in cost of production. At the present time, spirulina is commercially cultivated in several countries (Mexico, USA, Taiwan, Thailand, Japan, Israel), with a total production of a few hundred tons per annum.

Vonshak [20] in 1988 reported on main production sites (table 1). These products are sold in industrialized countries in the form of pills and spray-dried powder on the health food market. People, particularly vegetarians, looking for natural products and health-food supplements for their diet are attracted by spirulina's overall nutritional qualities.

Chemical Composition

The percent contribution of the major component of *S. platensis* and *S. maxima*, is shown in table 2. The proportion of the individual components varies to a certain extent due to the particular culture conditions.

Protein represents more than 60% and in certain samples even 70% of the dry weight. Spirulina represents one of the richest protein sources of plant origin, far more than meat and fish [15–25%] or soybean meal (35%). But it should be noted that approximately 15% of the crude protein (N × 6.25) is derived from nonprotein nitrogen. Fat content is 5–7% [25, 26]. Carbohydrates, accounting for 10–15% of the dry weight, are represented by a branched polysaccharide structurally similar to glycogen [27]. Another glucose-containing polysaccharide was also characterized [28] and accounts for 1% of the dry weight. Total nucleic acid content is less than 5% of the dry weight, which is lower than that of bacteria or yeasts in which it accounts for 4–10%. RNA has been reported to represent 2.2–3.5% of

Table 1. Spirulina production plants [from Vonshak and Richmond, 19]

Company	Location	Annual production tons dry weight
Sosa Texcoco	Mexico	300
Earthrise Farms	California, USA	90
Siam Algae	Thailand	100
Nippon-Spirulina	Japan	30
Cyanotech	Hawaii	40

Table 2. Approximate composition (% dry weight) of *S. platensis, S. maxima* and soybean meal

Sample	Water	Ash	Crude protein	Lipids	Carbo-hydrates	Crude fiber
S. platensis[1]	9	10	62	3.9	8.5	3
S. maxima[2]	4–7	6–9	60–71	4	8–13	1
Soybean	7–10	4	34–40	16–20	19–35	3–5

[1] Sun-dried. Data from Becker [25].
[2] From Lake Texcoco. Spray-dried. Data from Durand-Chastel [11].

the dry weight, whereas DNA represents 0.6–1% [26]. Thus, whereas *Candida lipolytica* grown on alkanes contains at least 1 g nucleic acid per 10 g of single cell protein, Spirulina contains only 0.6–0.7 g per 10 g of protein [10].

Spirulina can be considered a good source of vitamin B, especially B_{12}, but, it is less rich in vitamins B than yeast. Due to its high carotenoid [29] content it is a good source of β-carotene.

Composition and Nutritional Value of Proteins

The nutritional value of a protein depends on its amino acid composition and on the digestibility and bioavailibility of its essential amino acids.

Amino Acid Composition. Its amino acid composition has been reported by a large number of investigators [11, 25, 31–33]. Some representative values for these amino acids are presented in table 3.

Table 3. Data on amino acid composition of spirulina protein (mg/g crude protein N × 6.25) in comparison with FAO/WHO/UNU suggested pattern for preschool children

Amino acid	*S. platensis* [25]	*S. maxima* [31]	*S. maxima* [32]	FAO/WHO/UNU [34]
Histidine	22		17	19
Isoleucine	67	18	60	28
Leucine	98	60	87	66
Valine	71	80	63	35
Phenylalanine	53	50	49	
Tyrosine	53	40	40	63
Lysine	48	46	41	58
Methionine	25	14	20	25
Cystine	9	4		
Tryptophan	3	14	12	11
Threonine	62	46	49	34
Alanine	95	68	77	
Arginine	73	65	72	
Aspartic acid	118	86	99	
Glutamic acid	103	126	135	
Glycine	57	48	47	
Proline	42	39	39	
Serine	51	42	45	

Table 4. Comparative nutritional studies with spirulina (at the 10% protein level)

Protein source	Protein efficiency ratio	Biological value	Net protein utilization	Reference
S. maxima raw	–	63-0	47.7	31
S. platensis sun-dried	1.80	77.6	65.0	25
S. maxima spray-dried	2.18	–	56.6	39
S. maxima spray-dried	2.20	–	57.0	11

The essential amino acids comprise 47% of the protein. As table 3 shows, the content of lysine and sulfur-containing amino acid in spirulina is slightly lower than the FAO/WHO/UNU recommended pattern for preschool children [34].

The drying process has a major influence on the level of methionine: drum-drying reduces the methionine content by 30% by comparison with spray-drying.

This methionine deficiency, which is common in all types of single-cell protein can be compensated for by blending the algae with other conventional protein sources [25].

Due to the low level of reducing sugars in the alga, lysine is relatively stable during heat treatment [33].

Protein Digestibility. Owing to the absence of cellulose in its cell walls [28], spirulina does not require chemical or physical processing steps to make it digestible. In vitro, the fresh algal samples were more digestible (85%) than sun-dried or freeze-dried samples (70%) [36].

In rats, Durand-Chastel [11] reported an apparent digestibility of 85% with spirulina meal produced in Mexico. Vermorel et al. [37] reported a true digestibility of 73–79%.

The digestibility coefficient of raw spirulina was 74–76% [31]; it increased to 91–95% after sun-drying [25]. Sautier and Trémolières [38] studied the apparent nitrogen digestibility in 5 malnourished adults fed high-dose of spirulina (80–90 g/day) via gastric tube and found an average 90% absorption of nitrogen.

Nutritional Value of Spirulina Proteins. Unconventional sources of proteins such as microalgae have been subjected to detailed evaluation regarding their nutritional quality. The results of some rat feeding experiments: protein efficiency ratio, biological value, net protein utilization are summarized in table 4.

Differences in the data might be due to diverse processing methods or different strains used. Supplementation with methionine improved the protein efficiency ratio only marginally [25]. A number of nutritional studies have been performed in different animals (mice, rats, pigs, chicken, calves) fed diets in which spirulina was substituted totally or partially for the protein requirement [31, 37–41]. *S. maxima* and *S. platensis* either drum- or spray-dried are well accepted by various animals. Nitrogen deposition was comparable to that obtained with most other plant protein sources, if not better, but inferior to the best protein source lactalbumin. By adding methionine to the spirulina diet, its nutritional quality improved, although it never reached that of casein [25, 37].

There are few and incomplete reports of experiments in humans. Galvan [40] studied the nitrogen balance in 10 malnourished children aged 5–12 months who were fed in successive 4-day periods 2–3 g/kg body weight of spirulina, milk or soy milk. In spite of a lower protein digestibility (spirulina 60%, soya 70%) nitrogen retention was higher with spirulina (40%) than with soya (30%).

Sautier and Trémolières [38] administered successively 20, 50 and 100 g spirulina protein for 4–6 days to malnourished adults and measured nitrogen balance. Serum uric acid level did not increase during the experimental period.

In summary, spirulina has an extremely high protein content (up to 70% of dry weight) and its quality is among the best in the plant world.

The amino acid score corrected for digestibility procedure recommended recently by a Joint FAO/WHO Expert Consultation [42] as a measure of protein quality could be calculated for spirulina protein produced in India [25] and would give a score of 62: (1) true digestibility = 75% and (2) amino acid score = 82% (lysine is the first limiting essential amino acid with an amino acid ratio of 82%), and (3) score adjusted for digestibility = 82% × 75% = 62%.

Composition and Nutritional Value of Lipids

Composition. Glycolipids are the major lipid constituents of spirulina membranes [43–45]. They contain the sugar *D*-galactose as the polar moiety. The commonest glycolipids are monogalactosyldiacylglycerol (MGDG): they represent 70–80% of the total lipids. Another glycolipid is sulfoquinovosyldiacylglycerol (SQDG). Quinovose is an isomer of glucose (fig. 1).

Cyanobacteria (other than spirulina) have been shown to contain traces of another glycolipid: monoglycosyldiacylglycerol (MGlDG) [46] that could act as a precursor in galactolipid synthesis, glucose being isomerized into galactose [47] in the course of galactolipid synthesis. This remains to be demonstrated in spirulina. Only one type of phospholipid is present: small amounts of phosphatidylglycerol (PG).

Other phospholipids usually present in cellular membranes such as phosphatidylcholine (lecithin), phosphatidylethanolamine (cephalin), phosphatidylinositol, diphosphatidylglycerol (cardiolipin), or phosphatidylserine are absent.

The total fatty acid composition of spirulina membranes has been known for years [43–45, 48, 49, 51]. However, the precise composition of the polar head group and the positional distribution of fatty acids in the glycolipid molecules has been worked out in only a few cyanobacteria [47, 52].

A detailed analysis of spirulina glycolipids has been carried out in the laboratory of one of us (J.P.D.).

ω6 and ω3 are the two classes of essential polyunsaturated fatty acids defined by the position of the double bond closest to the methyl end of the molecule. ω6 is represented by 18:2ω6, linoleic acid (LA), and ω3 is represented by 18:3ω3, α-linolenic acid (LNA). In the human, both LA and LNA must come from the diet since the synthesis of these ω3 and ω6 structures has not been detected in animals. LA is universally found in the vegetable kingdom and is particularly rich in most, but not all, vegetable seeds and in the oils produced from seeds (with coconut oil, cocoa butter, and palm oil being exceptions). LNA, on the other hand, is more restricted in nature than LA, and is found in the chloroplast of green leafy plants and a few vegetable oils such as linseed, rapeseed, soybean and walnut oil.

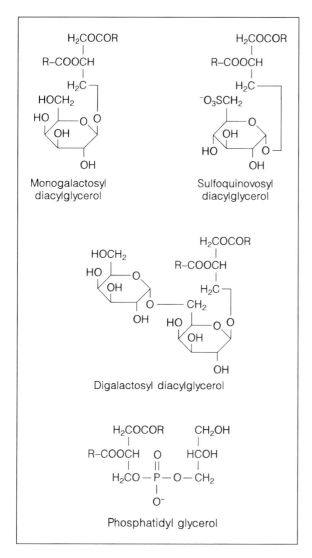

Fig. 1. Lipids in spirulina membranes.

Table 5 reports the fatty acid composition of various algae related to spirulina.

S. platensis is unique in that it contains exclusively γ-linolenic acid (GLA; 18:3ω6) instead of LNA (18:3ω3). It is the most concentrated source of GLA in the vegetable kingdom (1% of dry weight): 25–30% of fatty acids are 18:3ω6.

Table 5. Fatty acid composition (% dry weight) of various cyanobacteriae [from Kenyon et al., 50]

	Sat.	16:1	18:1	16:2	18:2	18:3ω3	18:3ω6
Anabaena cylindrica	50	6	6	6	24	11	0
Anabaena flos-aquas	41	6	5	4	37	11	0
Anabaena variabilis	39	15	14	0	14	17	0
Chlorogloca fritschii							
Photoautotrophic	44	17	14		13	12	
Photoheterotrophic	41	19	26	0	13	0	
Hapalosiphon laminosus	57	24	18	0	0	0	0
Oscillatoria williamsii	42	24	11	14	4	0	0
Spirulina platensis	46	10	5	0	12	0	21

Other vegetable sources such as evening primrose or black currant berry contain only 10–15%. Some other cyanobacteria also contain traces of GLA, but they are mixed with fairly large quantities of 18:3ω3 and 18:4ω6 [54].

It is worth noting that certain strains of spirulina have a different composition, as indicated in table 6.

Looking at the fatty acid composition of the various galactolipids classified according to the nature of their polar head, one can conclude that galactolipids are more unsaturated than PG or sulfolipid. As indicated in table 7, MGDG contains almost 40% of GLA, whereas PG has less than 5%.

Nutritional Value of Spirulina Lipids. From a nutritional and medical point of view, two lipid fractions present in spirulina are of interest: GLA and the sulfolipid fraction.

GLA is a precursor of dihomo-GLA and the prostaglandin PGE$_1$ series. It has been demonstrated that GLA as well as dihomo-GLA have beneficial effects in humans as well as laboratory animals, particularly on the vascular system [55].

A sulfolipid fraction has been recently claimed to inhibit in vitro the development of the AIDS virus [56]. Although sulfolipids are present in leaf chloroplasts (5–10% of chloroplast lipids, i.e. 2–5% of total lipids), this fraction is mixed with many other lipids from which it is difficult to separate. In cyanobacteria, on a quantitative scale, the sulfolipid is much more concentrated and in the absence of phospholipids much easier to separate. This makes spirulina a good candidate for preparing purified sulfolipid should this initial report of its efficacy on the AIDS virus be confirmed.

Table 6. Variability in fatty acid composition among 3 spirulina strains [from Kenyon et al., 50]

Strain	14:0	16:0	18:0	14:1	16:1	18:1	18:2	18:3 ω3	18:3 ω6
6313	21	33	1	12	27	2	–	–	–
H-1	25	11	2	14	34	2	–	–	8
7106	tr	33	4	–	2	7	12	–	27

Table 7. Distribution of fatty acids in various lipid fractions (% of total fatty acids) [pers. data]

	C16:0	C16:1	C18:0	C128:1	C18:2	C18:3ω6
Spirulina platensis 8005						
MGDG	38.84	8.38	10.25	5.25	2.42	34.85
PG	35.65	3.17	19.18	11.63	21.45	8.91
DGDG	23.37	7.92	31.29	12.48	3.96	20.99
SQ	26.87	2.29	42.29	9.64	11.93	6.99
Spirulina maxima						
MGDG	24.69	4.25	35.49	13.72	1.59	20.27
PG	28.71	3.94	42.59	12.95	–	11.82
DGDG	13.97	4.37	47.42	16.81	–	17.03
SQ	20.16	–	36.62	16.05	8.23	18.93
Spirulina platensis (from Vietnam)						
MGDG	39.47	8.38	6.91	4.47	2.05	38.72
PG	38.54	–	22.46	5.49	27.63	5.87
DGDG	32.79	5.88	20.19	7.68	3.22	30.24
SQ	41.01	2.43	17.87	6.63	24.20	7.87

Vitamins

Spirulina contains several pigments, among them large quantities of β-carotene as well as vitamin B, especially B_{12}. The pigment composition has been determined as chlorophyll a, β-carotene, echinenone, β-cryptoxanthin, zeaxanthine, myxoxanthophyll and oscilloxanthine [25, 57]. β-Carotene represents 0.15–0.20% dry weight of spray-dried spirulina and xanthophyll about 0.15% [26, 58].

Table 8. Vitamin B content of various spirulina (mg/100 g dry wt)

Spirulina	B_1	B_2	Niacin	Panto-thenic acid	B_6	B_{12}	Folates
Lab. culture [57]	4.1	3.6	9.8	0.15	0.32	0.13	0.034
Sun-dried [25]	2.8	3.3	–	–	0.13	0.24	–
Spray-dried [61]	3.9	4.0	10.7	0.11	0.30	0.28	0.056
Spray-dried [41]	3.9	4.5	9.2	2.5	0.8	0.23	–
Lyophilized [61]	4.2	3.5	10.7	0.15	0.32	0.13	–

Provitamin A (β-Carotene). Raw spirulina contains 700–1,700 µg of β-carotene per gram fresh weight [41, 59, 60]. A mean value of 1,400 µg would provide 220 RE/g of spirulina, which makes it a very good source of provitamin A. Compared to spray-drying, roller-drying decreases the content of β-carotene by about one third [41] and sun-drying by about a half [57].

In countries in which vitamin A deficiency and xerophthalmia are prevalent such as Bangladesh and Vietnam, locally grown spirulina could be considered as a safe and convenient substitute for other vitamin A supplementation.

Vitamin B. Most of the vitamins have been found in spirulina, and their concentration has been determined [25, 26, 31, 41]. The vitamin B content of various sources of spirulina appears in table 8. The levels of B_1, B_2, B_6 and niacin (vitamin PP) are comparable to the level found in yeast. Besides, spirulina is the richest nonanimal source of cobalamine.

Minerals

Mineral content varies depending on the composition of the medium in which spirulina is grown; therefore, in table 9, the values are expressed as a concentration range [11].

Of special interest is iron, whose mean concentration is 0.5 g/kg. Johnson and Shubert [62] recently investigated the bioavailability of iron in rats from cultured and commercially available spirulina. The data demonstrated that the cultured spirulina was as good a source of iron as iron sulfate.

These authors also pointed out that both cultured and commercial spirulina contained approximately 9.5 ppm mercury 'so that chronic use may lead to mercury intake above prudent level'. That algae can absorb trace elements while growing has been confirmed by Godinez at al. [63]. The heavy metals in algae may be derived from the water and fertilizer used for cultivation. More recently, Slotton et al. [64], using another analytical technique for mercury and lead detection,

Table 9. Mineral content (range mg/kg dry wt) in spirulina [from Durand-Chastel, 11]

Potassium	13,300	15,400
Phosphorus	7,600	8,900
Chloride	4,000	4,400
Magnesium	1,400	1,900
Calcium	1,040	1,300
Iron	470	570
Sodium	270	410
Zinc	27	39
Manganese	18	25

reported that the concentration of these toxic elements in commercially grown spirulina is such that it would take 77 g of the algal supplement per day to reach the daily intake caution guideline levels of the FAO/WHO. In the 1970s, spirulina underwent extensive safety studies. Independent testing in France [65], Mexico and Japan showed no toxic side effects on humans, rats, pigs, chickens and fish.

In 1980, the United Nations Industrial Development Organization (UNIDO) sponsored a large and comprehensive toxicity study on rats and mice [66]. No second- or third-generation reproduction, fertility, lactation or birth defects were found. The study demonstrated its complete safety as a human food.

Spirulina as a Human Food

Spirulina has a mild marine odor that is stronger than the taste. In regions where the population is not familiar with spirulina, as well as in the preparation of baby food, the color may represent a problem: in such instances, it would be convenient to decolorize the product, which can be done by extracting the pigments with suitable solvents.

In developing countries there is a need for small production units at village level, using local resources [67] to supply protein and β-carotene for human and animal (including fish) consumption. Carbon dioxide from biogas digesters fueled by plant, animal or human wastes can be recycled to grow spirulina. On the village level, this has been achieved by an integrated system [68] now in operation in Togo, Peru, India and Vietnam.

Today, while commercial farms grow spirulina as a health food, village-scale projects in Africa, Asia and South America can produce spirulina for local consumption and help fight protein and vitamin A malnutrition.

Conclusion

Algae have long been praised as one of the most promising sources of protein of the future. So far, food shortages have not been so pressing that algae have become a desirable replacement for animal protein. However, population pressure in certain areas of the world could quickly change the picture. The simple cultivation technology as well as the good quality of its protein, the high level of β-carotene and the absence of any side effects favor the large-scale production of spirulina in rural areas of the developing world.

In industrialized countries, quite apart from considerations of enrichment with protein or other nutrients, the presence of GLA in large quantities in spirulina may contribute to the prevention of cardiovascular disorders.

References

1   Farrar W: Tecuitlal: A glimpse of Aztec food technology. Nature 1966;211:341–342.
2   Brandilly M: Depuis des lustres une tribu primitive du Tchad exploite la nourriture de l'an 2000. Sci Avenir 1959;152:516–519.
3   Dangeard P: Sur une algue bleue alimentaire pour l'homme, *Arthrospira platensis* (Nordst) Gom. Actes Soc Linn (Bordeaux) 1940;91:39–41.
4   Delpeuch F, Joseph A, Cavalier C: Consommation alimentaire et apport nutritionnel des algues bleues *(Oscillatoria platensis)* chez quelques populations du Kanem (Tchad). Ann Nutr Aliment 1975;29:497–516.
5   Léonard J, Compère P: *Spirulina platensis* (Gom) Geitler, algue bleue de grande valeur alimentaire par sa richesse en protéines. Bull Jard Bot Natl Belg 1967;37(suppl 1).
6   Clément G: Comment: A new type of food algae; in Mateles R, Tannenbaum S (eds): Single Cell Protein. Cambridge, MIT Press, 1968, pp 306–308.
7   Clément G: Production et constituants caractéristiques des algues *Spirulina platensis et maxima*. Ann Nutr Aliment 1975;29:477–478.
8   Clément G: Producing Spirulina with CO_2; in Tannenbaum S, Wang D (eds): Single Cell Protein. II. Cambridge, MIT Press, 1975, pp 467–474.
9   Meyer C: Etude d'une culture d'algues en vue d'une production à grande échelle. Ind Aliment Agric 1969;86:1445–1449.
10 Durand-Chastel H, Clément G: Spirulina algae: Food for tomorrow. Proc 9th Int Congr Nutrition, Mexico, 1972. Basel, Karger 1975, vol. 3, pp 85–90.
11 Durand-Chastel H: Production and use of spirulina in Mexico; in Shelef G, Soeder CJ (eds): Algae biomass. Amsterdam, Elsevier/North Holland, 1980, pp 51–63.
12 Clément G, Rebeller M, Trambouze P: Utilisation massive de gaz carbonique dans la culture d'une nouvelle algue alimentaire. Rev Inst Fr Pétroles 1968;23:702–711.
13 Zarrouk C: Contribution à l'étude d'une Cyanophycée. Influence de divers facteurs physiques et chimiques sur la croissance et la photosynthèse de *Spirulina maxima* (Setch et Gardner) Geitler; thèse sciences, Paris, 1966.
14 Kosaric N, Nguyen H, Bergougnou M: Growth of *Spirulina maxima* algae in effluents from secondary waste-water treatment plants. Biotechnol Bioeng 1974;16:881–896.
15 Nguyen H, Kosaric N, Bergougnou M: Some nutritional characteristics of *Spirulina maxima* algae grown in effluents from biological treatment plant. Can Inst Food Sci Technol J 1974;7:114–116.
16 Cifferi O: Spirulina, the edible microorganism. Microbiol Rev 1983;47:551–578.
17 Cifferi O, Tiboni O: The biochemistry and industrial potential of spirulina. Annu Rev Microbiol 1985;39:503–526.

18   Richmond A (ed): Handbook of Microalgal Mass Culture. Boca Raton, CRC Press, 1986.
19   Vonshak A: Recent advances in microalgal biotechnology. Biotech Adv 1991;8:709–727.
20   Vonshak A, Richmond A: Mass production of the blue-green alga spirulina. An overview. Biomass 1988;15:233–247.
21   Fox R: Algoculture: La spiruline, un espoir pour le monde de la faim. Aix-en-Provence, Edisud, 1986.
22   Faucher O, Coupal B, Leduy A: Utilization of seawater-urea as a culture medium for *Spirulina maxima.* Can J Microbiol 1979;25:752–759.
23   Materassi R, Tredici M, Balloni W: Spirulina culture in sea water. Appl Microbiol Biotechnol 1984; 19:384–386.
24   Jassby A: Spirulina: A model for microalgae as human food; in Lembi CA, Waaland JR (eds): in Algae and Human Affairs. Cambridge, Cambridge University Press, 1988.
25   Becker EW: Production and utilization of the blue-green alga spirulina in India. Biomass 1984;4: 105–125.
26   Santillan C: Mass production of spirulina. Experientia 1982;38:40–43.
27   Casu B, Naggi A, Vercellotti J: Polisaccardi di reserva della *Spirulina platensis*; in Materassi R (ed): Prospettive della coltura di spirulina in Italia. Consiglio Nazionale della Ricerca, Rome, 1980.
28   Van Eykelenburg C: On the morphology and ultrastructure of the cell wall of *Spirulina platensis.* Ant Van Leeuwenhoek J Microbiol Serol 1977;43:89–99.
29   Dang DK: The pigment composition of spirulina. Research Seminar and Workshop on mass cultures of microalgae, Silpakorn University, 1991.
30   Dillon JC, Young V, Scrimshaw NS: Single-cell protein: Utilization in human feeding. Deuxième Symposium International Alimentation et Travail. Paris, Masson, 1974, pp 459–472.
31   Clément G, Giddey C, Menzi R: Amino acid composition and nutritive value of the alga *Spirulina maxima.* J Sci Food Agric 1967;18:497–501.
32   Wu J, Pond W: Amino acid composition and microbial contamination of *Spirulina maxima,* a blue-green alga, grown on the effluent of different fermented animal wastes. Bull Environ Cont Toxicol 1981;27:361–363.
33   FAO: Nutritional Studies. Amino acid content of foods and biological data on proteins. Rome, FAO, 1970, No 24.
34   Joint FAO/WHO/UNU Expert consultation: Energy and Protein Requirements. Techn Rep Ser WHO 1985;724.
35   Adrian J, Frangne R: Comportement thermique des microorganismes alimentaires: Levures et algues spirulines. Ind Aliment Agric 1975;92:1365–1375.
36   Anasuya Devi M, Subbulakshmi G, Mahdavi Devi, Vankataraman L: Studies on the proteins of mass-cultivated, blue-green alga *(Spirulina platensis).* J Agric Food Chem 1981;29:522–525.
37   Vermorel M, Toullenc G, Dumond D, Pion R: Valeur protéinique et énergétique des algues bleues spirulines supplementées en acides aminés: Utilisation digestive et métabolique par le rat en croissance. Ann Nutr Aliment 1975;29:535–552.
38   Sautier C, Trémolières J: Valeur alimentaire des algues spirulines chez l'homme. Ann Nutr Aliment 1975;29:517–533.
39   Bourges H, Sotomayor A, Mendoza E, Chavez A: Utilization of the alga spirulina as a source of protein. Nutr Rep Int 1971;4:31–43.
40   Galvan R: Experimentacion clinica con espirulina. Colloque sur la valeur nutritionnelle des algues spirulines, Rueil, May 1973.
41   Bujard E, Bracco U, Mauron J, Mottu F, Nabholz A, Wuhrmann J, Clément G: Composition and nutritive value of blue-green algae (spirulina) and their possible use in food formulations. 3rd Int Congr Food Sci Technol, Washington, 1970.
42   Protein quality evaluation. Report of a Joint FAO/WHO Expert. Rome, FAO/WHO, 1990.
43   Nichols B: Lipid metabolism; in Carr N, Whitton B (eds): The biology of Blue-Green Algae. Oxford, Blackwell, 1973.
44   Nichols B, Wood B: The occurrence and biosynthesis of gamma linolenic acid in blue green algae: *Spirulina platensis.* Lipids 1968;3:46–50.
45   Erwin JA: Comparative biochemistry of fatty acids in eukaryotic microorganisms; in Erwin JA (ed): Biomembranes of Eukaryotic Microorganisms. London, Academic Press, 1973, pp 42–145.

46   Fiege G, Heinz E, Wrage K, Cohems N, Ponzelar E: Discovery of a new glyceroglycolipid in blue-green algae and its role in galactolipid biosynthesis; in Mazliak P, Benveniste P, Costes C, Douce R (eds): in Biogenesis and Function of Plant Lipids. Amsterstam, Elsevier, 1978, pp 135–143.

47   Sato N, Murata N: Lipid biosynthesis in the blue-green algae, *Anabaena variabilis*. II. Fatty acids an lipid molecular species. Biochim Biophys Acta 1982;710:279–289.

48   Cohen Z, Vonshak A, Richmond A: Fatty acid composition of spirulina strains under various environmental conditions. Phytochemistry 1987;26:2225–2228.

49   Nichols B, Harris R, James A: The lipid metabolism of blue-green algae. Biochem Biophys Res Commun 1965;230:256–262.

50   Kenyon C, Rippka R, Stanier R: Fatty acid composition and physiological properties of some filamentous blue-green algae. Arch Mikrobiol 1972;83:216–236.

51   Stanier R, Cohen-Bazire G: Phototrophic prokaryotes: The cyanobacteria. Ann Rev Microbiol 1977;31:225–274.

52   Zepke H, Heinz E, Radunz A, Linnscheid M, Pesh R: Combination and positional distribution of fatty acids in lipids from blue-green algae. Arch Mikrobiol 1978;119:157–162.

53   Mazliak P: Le métabolisme des lipides dans les plantes supérieures. Paris, Masson, 1968.

54   Canto de Loura I, Dubacq JP, Thomas JC: The effects of nitrogen deficiency on pigments and lipids of cyanobacteria. Plant Physiol 1987;83:838–843.

55   Dillon JC: Essential fatty acid metabolism in the elderly: Effects of dietary manipulation; in Horrisberger M, Bracco U (eds): Lipids in Modern Nutrition. New York, Nestlé Nutrition/Raven Press, 1987, pp 93–106.

56   Gustafson K, Cardelina J, Fuller R, Weishow O, Kiser R, Snader K, Patterson G, Boyd M: AIDS antiviral sulfolipids from cyanobacteria. J Natl Cancer Inst 1989;81:1254–1258.

57   Busson F: Etude de *Spirulina platensis* (Gom.) Geitler et de *Spirulina geitleri* J de Toni, Cyanophycées alimentaires; thèse doctorat pharmacie, Marseille, 1971.

58   Paoletti C, Pushparaj B, Florenzano G, Capella P, Lerker G: Unsaponifiable matter of green and blue-green algae lipids as a factor of biochemical differentiation of their biomasses. I. Total unsaponifiable and hydrocarbon fraction. Lipids 1976;11:258–265.

59   Palla JC, Mille G, Busson F: Etude comparée des caroténoïdes de *Spirulina platensis* (Gom.) Geitler et de *Spirulina geitleri* J. de Toni (Cyanophycées) CR Acad Sci Sér D 1970;270:1038–1041.

60   Palla JC, Busson F: Etude des caroténoïdes de *Spirulina platensis* (Gom.) CR Acad Sci Sér D 1969; 269:1704–1707.

61   Pfrommer A, Slump P, Willems J: Chemical and biological evaluation of two samples of algae. Rep No R 3193. TNO, 1970.

62   Johnson P, Shubert E: Availability of iron to rats from spirulina, a blue-green algae. Nutr Res 1986; 6:85–94.

63   Godinez JL, Ortega MM, De La Lanza E: Study of the edible algae of the valley of Mexico. IV. Analysis of some inorganic elements. Nutr Rep Int 1984;30:1279–1285.

64   Slotton D, Goldman C, Franke A: Commercially grown spirulina found to contain low levels of mercury and lead. Nutr Rep Int 1989;40:1165–1172.

65   Boudene C, Collas E, Jenkins C: Recherche et dosage de divers toxiques minéraux dans les algues spirulines de différentes origines, et évaluation de la toxicité à long terme chez le rat d'un lot d'algues spirulines de provenance mexicaine. Ann Nutr Aliment 1975;30:577–588.

66   Chamorro-Cevallos G: Etude toxicologique de l'algue spiruline. UNIDO/ 10.387. UF/MEX/78/ 048, 1990.

67   Dank DK: The outdoor mass culture of *Spirulina platensis* in Viet-Nam. J Appl Physiol 1990;2: 179–181.

68   Fox R: Spirulina, the alga that can end malnutrition. Futurist 1985;19:30–35.

J.C. Dillon, Institut National Agronomique, Nutrition Humaine, 16, rue Claude-Bernard,
F–75005 Paris (France)

Simopoulos AP (ed): Plants in Human Nutrition.
World Rev Nutr Diet. Basel, Karger, 1995, vol 77, pp 47–74

Purslane in Human Nutrition and Its Potential for World Agriculture

Artemis P. Simopoulos[a], Helen A. Norman[b], James E. Gillaspy[c]

[a] The Center for Genetics, Nutrition and Health, Washington, D.C.,
[b] Weed Science Laboratory, US Department of Agriculture-Agricultural Research
   Service, Beltsville, Md.,
[c] Austin, Tex., USA

Contents

Introduction 47
Geographic Distribution and Origin 50
Food and Agricultural Potential of Purslane 51
    Purslane Leaf Content of ω3 Fatty Acids, Antioxidant Vitamins
        (C, E, β-Carotene) and Glutathione 54
        Fatty Acid Content 54
        Antioxidant Content 56
    Profiles of Leaf ω3 Fatty Acids and Antioxidants throughout Plant Development
        in Growth Chamber Conditions 60
    Agricultural Potential of Purslane 64
Medicinal Uses of Purslane 67
Contribution of Purslane to Food Technology: Pectin 69
Conclusions 70
References 71

> I have made a satisfactory dinner ...
> off a dish of purslane *(Portulaca oleracea)*
> which I gathered in my cornfield ...
> *Henry David Thoreau:* Walden, 1854

Introduction

Purslane is the common name for certain small fleshy annual plants of the genus *Portulaca,* with prostrate, reddish stems, egg-shaped leaves attached by their narrowest end, and small yellow flowers that open in the sunlight. The

Fig. 1. Wild purslane *(Portulaca oleracea).*

botanical name 'portulaca' chosen for this family refers to the juicy qualities of its members, which carry (porto) milk or soup (lac). The common name 'purslane' is a corruption from the French 'pourpier'. There are at least 100 *Portulaca* species. The one we are interested in is the wild form of *Portulaca oleracea* (fig. 1).

We have previously shown that purslane is an excellent source of α-linolenic acid (LNA, 18:3ω3) (table 1) and antioxidants [1, 2]. One hundred grams of fresh purslane leaves (one serving) contain about 300–400 mg of 18:3ω3; 12.2 mg of α-tocopherol; 26.6 mg of ascorbic acid; 1.9 mg of β-carotene; and 14.8 mg of glutathione [2]. In fact, purslane is the richest source of LNA, richer than any other green leafy vegetable investigated to date. ω3 fatty acids (FA) are essential FA for growth and development, and have antithrombotic, hypolipidemic, hypotensive and anti-inflammatory properties. Extensive research in the past 10 years on LNA and the longer-chain FA, eicosapentaenoic acid (EPA) and docosahexaenoic acid (DHA) found in fish and fish oils, has emphasized the important role of ω3 FA in growth and development and in health and disease [3–8]. International epidemiologic studies on the relationship between diet and coronary artery disease and cancer suggest that populations with higher intakes of fruits and vegetables have less coronary artery disease and cancer [9, 10]. In interpreting the stud-

Table 1. Fatty acid content of plants[1]

Fatty acid	Purslane	Spinach	Buttercrunch lettuce	Red leaf lettuce	Mustard
14:0	0.16	0.03	0.01	0.03	0.02
16:0	0.81	0.16	0.07	0.10	0.13
18:0	0.20	0.01	0.02	0.01	0.02
18:1ω9	0.43	0.04	0.03	0.01	0.01
18:2ω6 (LA)	0.89	0.14	0.10	0.12	0.12
18:3ω3 (LNA)	4.05	0.89	0.26	0.31	0.48
20:5ω3 (EPA)	0.01	0.00	0.00	0.00	0.00
22:6ω3 (DHA)	0.00	0.00	0.001	0.002	0.001
Other	1.95	0.43	0.11	0.12	0.32
Total fatty acid content	8.50	1.70	0.60	0.702	1.101

Values are expressed as mg/g of wet weight.
[1] Modified from Simopoulos and Salem [1].

ies, 'protective factors' were searched for and it was concluded that the fiber content of fruits and vegetables was responsible for the lower mortality rates from coronary artery disease and cancer, which led to government recommendations to increase dietary intake of fiber [9, 10]. Little attention was paid to the ω3 FA content or the antioxidant content of vitamins and glutathione until recently [11].

Studies reviewed by Rifici and Khachadurian [12] have shown that vitamin E, vitamin C and β-carotene taken as supplements decrease the rate of oxidation of low-density lipoproteins (LDL) and therefore the risk for deposition of LDL in the atheroma, which leads to atherosclerosis. Studies on paleolithic nutrition by Eaton and Konner [13] suggest that during human evolution, the diet was rich in ω3 fatty acids and antioxidants. Our studies on the composition of purslane support their suggestion that wild plants are rich in vitamin C and that paleolithic humans had an average vitamin C intake from wild plants and fruits of about 390 mg/day, instead of the 88 mg/day available in the US diet or the 60 mg/day recommended by the US Recommended Dietary Allowances.

In addition to being a rich source of LNA, vitamin E and glutathione, purslane is also a rich source of pectin. Although its unique nutritional composition was not known thousands of years ago, purslane was recognized as an important plant to human health by many peoples around the world, both for its nutritional and medicinal properties. This paper presents information on the geographic distribution and origin of purslane; its food and agricultural potential; the purslane leaf content of ω3 FA, antioxidant vitamins and glutathione; profiles of leaf ω3 FA and antioxidants throughout plant development in growth chamber conditions; medicinal uses of purslane and studies on pharmacologic aspects; and the contribution of purslane to food technology, including its pectin content.

Geographic Distribution and Origin

Coquillat [14] rated purslane as the eighth most common plant in the world, whereas Holm et al. [15] state that it is one of three most frequently reported weeds across the world, but consider it to be the ninth in troublesomeness to world agriculture. Authors commonly refer to purslane as having been brought to North America from Europe by man, and indeed such introductions have almost undoubtedly been made. However, archaeological data also indicate a pre-Columbian presence in North America [16–18]. Recently, the increased recovery of carbonized plant remains from archaelogical sites has begun to give new insight into prehistoric habitats, as well as human subsistance patterns [17]. Seeds identified as *P. oleracea* have been recovered from an archaeological site in Illinois (USA) [18]. Radiocarbon determinations place *P. oleracea* at Salts Cave, Kent., USA [19, 20] during the first millennium BC. It should be noted that *Portulaca* sp. was reported to be among the plant remains in the fire level at the bottom of the Great Mound at the Troyville site in Catahoula Parish, La., USA [21], which establishes the presence of these plants in south Louisiana around 500 AD.

Danin et al. [22] made chromosome counts and, by means of a dissecting microscope, studied the seeds of *P. oleracea*. They chose not to study what they termed a high diversity of forms in Australia and New Zealand, which differ from those in all other parts of the world. Within the area of coverage, they treated the species as a complex of nine diploid, tetraploid or hexaploid subspecies having either 18, 36, or 54 chromosomes.

Their summarization states:

'In our study we found that long distance dispersal by sea and the 'weedy' properties of *P. oleracea* could play a considerable role in its present day distribution. This makes the attempts to arrive at any conclusions regarding the origin of *P. oleracea* rather difficult. The highest number of subspecies and local forms occurs in Mexico. The presence of all three ploidy types there may suggest that this is at least the center of diversity of *Portulaca oleracea*' [22].

Table 2. Common names of *Portulaca oleracea* in some languages

Language	Common names
English	Purslane, Pursley, Pussley, Pusley
French	Pourpier
Australia	Low Pigweed
Spanish	Verdolaga
Greek	Andrachla, Glystrida
Italian	Portulaca
Philippino	Alusiman (local name)
(India)	Chhota, Ionia, Kurfa, Lonax, Munya

Food and Agricultural Potential of Purslane

It appears well established that purslane has served as human food since prehistoric times [16–18]. For the historic period, in addition to the writings of the ancient Greeks, the Oxford English Dictionary provides useful information. The description of purslane is given as: 'A low succulent herb, *Portulaca oleracea*, widely distributed throughout tropical and warmer temperate regions, used in salads, and sometimes as a potherb, or for pickling. Also called *common* or *garden purslane*. Formerly cultivated in English kitchen gardens, but now rarely met with.' As a possible index of the extent of earlier acquaintance with and usage of purslane and consequent effect on language, 29 variations in spelling are given, the oldest 'purcelan' in the year 1387 [23]. In a survey of 45 countries, Holm et al. [15] discovered 58 different names in use for purslane. Whether as food or as weed, mankind has evidently needed a word by which to discuss purslane (table 2).

Purslane has been and remains chiefly harvested from the wild or a product of the kitchen garden rather than an item of agriculture and commerce [24]. In Europe there has been sufficient interest in it that a large-leafed, erect garden cultivar has been developed and is commercially available [25]. Like spinach, it contains oxalic acid at levels not toxic for human consumption or for grazing animals that consume other plants as well. It is of wide prevalence and eaten readily by animals without apparent harm when part of a mixed diet, as that of animals in the wild or roaming free in natural areas. It is reported to be an excellent, highly relished feed for swine [26].

Thus its nutritional riches reached human beings through animal food as well as directly, prior to the day of feed lots and battery-reared poultry. Today purslane

Table 3. Purslane: nutritional composition[1] per 100 g

Protein	2.0 g	Water	91.2%
Calcium	79 mg	Energy	26 kcal
Phosphorus	32 mg	Thiamin	0.02 mg
Iron	3.6 mg	Riboflavin	0.10 mg
Carbohydrate	5.0 g	Niacin	0.5 mg
Fat	0.4 g	Vitamin A	260 ER
Fiber	0.9 g	Vitamin C	23 mg

[1] In Simopoulos [28].

Table 4. Amino acid composition of seeds

Amino acids	*P. oleracea*	Egg[1]	Soybean
Essential amino acids			
Iso	40	54	51
Leu	65	86	77
Lys	44	70	69
Met + Cys	27	57	16
Phe + Tyr	80	93	50
Thr	32	47	43
Val	49	66	54
Nonessential amino acids			
Ala	41	–	–
Arg	107	–	–
Asp	92	–	–
Glu	172	–	–
Gly	81	–	–
His	28	–	–
Pro	35	–	–
Ser	30	–	–
Total	923	–	–
Chemical score	40	100	74

Essential, nonessential, and total amino acids are expressed as mg/g of protein. Whole egg, with a chemical score of 100, is listed as a reference standard and the composition of soybean is included for comparison.

Modified from Miller et al. [29].

[1] FAO/WHO Ad Hoc Expert Committee, 1973.

Table 5. Amino acid composition of vegetative matter (leaves) of *P. oleracea*

Essential amino acids		Nonessential amino acids	
Iso	53/50	Ala	69/66
Leu	91/91	Arg	52/58
Lys	63/69	Asp	111/107
Met + Cys	13/14	Glu	132/127
Phe + Tyr	82/86	Gly	59/56
Thr	41/41	His	20/22
Val	69/66	Pro	46/45
		Ser	43/42
Total amino acids	943/940		
Chemical score	37/40		

Values are expressed as mg/g of protein.
Modified from Miller et al. [29].

is grown to some extent as a potherb, mostly in Europe. The *American Horticulturist* [27] states: 'Purslane may have been introduced to Massachusetts as a salad green or potherb by the colonists as early as 1672.' Interest in this herb continues today. Its mild flavor, palatability and mucilaginous qualities give it a wide range of uses as a kitchen vegetable. Stems and leaves can be eaten raw, alone or mixed with other greens. The plant can also be cooked like spinach. In his book *Walden,* Thoreau wrote of making a satisfactory dinner from a dish of purslane which he gathered and boiled. Purslane can be frozen like spinach or it can be dried and stored in jars for year-round use as a tasty cooked green with a flavor entirely different from that of fresh purslane.

As can be seen from table 3 [28], purslane is high in vitamin A, vitamin C, calcium, phosphorus and iron. Tables 4 and 5 [29] present information on the amino acid content of seeds and leaves. While low in methionine and cystine, purslane seeds have a good balance and a good concentration of essential amino acids with a sufficiently high chemical score [29]. The same is true about the amino acid composition of leaves. Purslane has a fat content of 15.9% of dry weight and a protein content of 12.8% which are of sufficient quality for use as a ruminant feed or as a grain lysine supplement [29].

In view of the wide distribution of purslane around the world and the archaeologic evidence of association with humans for thousands of years we have extended our studies on the nutritional composition of purslane. In this paper we

Table 6. Comparison of fatty acid profiles in total lipid extracts from leaves of purslane and spinach[1]

FA	Chamber-grown purslane		Wild purslane		Spinach	
	wt%	mg/g fresh wt	wt%	mg/g fresh wt	wt%	mg/g fresh wt
14:0	0.12	0.007	0.15	0.007	0.14	0.001
14:2	0.98	0.056	0.64	0.032	0.70	0.006
16:0	10.81	0.616	14.12	0.713	11.94	0.106
16:1 trans	2.46	0.140	2.92	0.147	2.32	0.021
16:2	ND	ND	ND	ND	1.91	0.017
16:3	ND	ND	ND	ND	14.62	0.130
18:0	1.12	0.064	0.95	0.048	0.78	0.007
18:1	4.99	0.016	2.13	0.108	2.04	0.018
18:2	16.99	0.968	13.45	0.704	11.70	0.104
18:3	59.87	3.412	63.78	3.221	53.85	0.480
20:0	1.46	0.083	0.94	0.047	ND	ND
24:0	1.20	0.068	0.92	0.046	ND	ND
Total content	5.43±0.25[2]		5.07±0.20		0.89±0.04	

ND = Not detected.

[1] Reproduced from Simopoulos et al. [2].

[2] Data represent mean values from four analyses each with three replicates per plant species/type.

present information on the ω3 FA content and distribution, and on the composition of purslane *(P. oleracea)* in terms of vitamin C, β-carotene, α-tocopherol and glutathione.

Purslane Leaf Content of ω3 FA, Antioxidant Vitamins (C, E, β-Carotene) and Glutathione

Fatty Acid Content

The growth conditions, plant materials and methods we have used have been published previously [2]. In table 6, the FA composition of total lipid extracts from leaves of two types of purslane plants (chamber grown and wild) is compared with equivalent extracts of spinach. The predominant FA in each extract was LNA (18:3ω3). The total FA content of chamber-grown purslane was about seven times higher than that found in spinach. The levels of 18:3ω3 in chamber-grown purslane were about seven times greater than spinach also. The wild purslane had

Table 7. Polar lipid content of chamber-grown purslane and spinach leaves[1]

Lipid	Percent of total polar lipids (mol)	
	purslane	spinach
Monogalactosyldiacylglycerol	48.7	44.2
Digalactosyldiacylglycerol	27.6	24.1
Sulfolipids	3.5	4.3
Phosphatidylcholine	7.5	11.2
Phosphatidylethanolamine	3.6	5.6
Phosphatidylglycerol	6.4	7.8
Phosphatidylinositol	1.5	1.7
Phosphatidylserine	1.2	1.0

[1] Reproduced from Simopoulos et al. [2].

some differences in relative amounts of individual FA, but the total weight of FA was not significantly different from that of plants grown from seed in an environmental growth chamber. Neither EPA nor DHA were present in these purslane samples, which is in contrast to other purslane plants analyzed [1, 30]. This may be due to a difference in cultivar, or due to environmental and developmental factors or plant growth stage. The addition of internal standards prior to lipid purification and FA methylation did not indicate that any selective losses of EPA or DHA due to oxidation occurred during the analysis. Levels of 18:3ω3 were at least tenfold lower in a previous report of purslane lipids [30]. The level of 18:3ω3 in cultivated purslane harvested from a different location was determined to be 3.8–4.2 mg/g fresh weight (data not shown).

The distribution of polar lipids in leaf extracts was similar for chamber-grown purslane and spinach (table 7), and typical for leaves of higher plants [31]. Analyses of the FA composition of major individual galactolipids and phospholipids recovered from purslane leaf extracts are shown in table 8, and a profile of the free FA pool in table 9. Monogalactosyldiacylglycerol was predominant in contributing approximately 2.0 mg of 18:3ω3/g fresh weight. Phosphatidylglycerol contained trans-Δ-3-hexadecaenoic acid (16:1 trans); this unusual FA is exclusively located at the *sn*-2 position of phosphatidylglycerol in chloroplasts [31], and is distinct from hydrogenation products found in some commercial vegetable oils. The free FA component was characteristically rich in saturated FA.

Table 8. Fatty acid composition of galactolipids and phospholipids from chamber-grown purslane leaves[1]

FA	Lipid class					
	MGDG	DGDG	SL	PC	PE	PG
14:0	0.11	0.17	0.19	1.25	0.97	0.16
14:2	0.30	0.48	1.21	1.01	2.38	1.25
16:0	4.36	15.37	25.70	19.31	25.10	32.26
16:1 trans	ND	ND	ND	ND	ND	15.27
18:0	0.75	0.47	2.16	1.64	1.97	1.24
18:1	1.23	5.08	7.43	7.27	8.00	8.00
18:2	7.83	11.96	15.31	21.16	19.50	13.44
18:3	85.42	65.97	48.00	48.36	42.08	28.38

MGDG = monogalactosyldiacylglycerol; DGDG = digalactosyldiacylgylcerol; SL = sulpholipids; PC = phosphatidylcholine; PE = phosphatidylethanolamine; PG = phosphatidylglycerol; ND = not detected.
[1] In Simopoulos et al. [2].

Table 9. Composition of chamber-grown purslane leaf free FA fraction[1]

FA	Percent of total wt
14:0	7.26
16:0	53.64
18:0	24.08
18:1	7.63
18:2	4.42
18:3	2.97
20:0	traces[2]
22:0	traces
24:0	traces

[1] In Simopoulos et al. [2].
[2] <1% of total wt.

Antioxidant Content

Wild plants are typically known to have higher levels of vitamin C than cultivated ones [13]. Studies from paleolithic nutrition indicate that the amount of vitamin C obtained by humans from eating a variety of wild plants was much higher, about 390 mg/day versus the 88 mg average intake obtained today in the US [13].

Table 10. Antioxidant content of purslane and spinach leaves[1]

	α-Tocopherol	Ascorbic acid	β-Carotene
Content, mg/100 g fresh weight			
Chamber-grown purslane	12.2±0.4	26.6±0.8	1.9±0.08
Wild purslane	8.2±0.3	23.0±0.6	2.2±0.1
Spinach	1.8±0.09	21.7±0.5	3.3±0.5
Content, mg/100 g dry weight			
Chamber-grown purslane	230±9	506±17	38.2±2.4
Wild purslane	170±8	451±14	43.5±3.0
Spinach	36±4	430±15	63.5±5.7

Data represent mean value from four analyses each with three replicates per species/type.

[1] In Simopoulos et al. [2].

We have determined levels of endogenous antioxidants (α-tocopherol, ascorbic acid, β-carotene and glutathione) in plant leaves sampled simultaneously for lipids as previously reported [2]. Table 10 shows the content of α-tocopherol, ascorbic acid and β-carotene in chamber-grown purslane, wild purslane and spinach expressed in mg/100 g fresh weight and in mg/100 g dry weight.

α-Tocopherol and Ascorbic Acid. In a previous survey of the antioxidant content of nine different weed species, it was recorded that levels of α-tocopherol ranged from 10 to 83 mg, and levels of ascorbic acid ranged from 2 to 861 mg/100 g dry weight (table 11) [32]. Relative to these findings, and other reports, levels of α-tocopherol found in purslane, 230 ± 9 mg/100 g dry weight, are up to ten times higher than has been recorded in other weeds (table 10 and 11) [2]. α-Tocopherol was present in spinach leaves at a level of 30–40 mg/100 g dry weight (table 10). The ascorbic acid content of chamber-grown purslane fell within the range previously reported for other weed species (table 11) [32], but was significantly higher (506 ± 17 mg/100 g dry weight) than the level found in spinach leaves (430 ± 5 mg/100 g dry weight) (table 10).

β-Carotene. In photosynthetic tissues of higher plants, β-carotene and other carotenoids are localized in chloroplasts; while there is little qualitative difference in the pigments present, there is considerable quantitative variation between different species [33, 34]. The levels of β-carotene were not significantly different in

Table 11. Antioxidant content of different plant species[1]

Plant species	Antioxidants, mg/100 g dry weight	
	ascorbic acid	α-tocopherol
Morning glory	2 ± 1	10 ± 3
Lamb's quarter	58 ± 18	12 ± 3
Alfalfa	143 ± 12	10 ± 1
Pigweed	504 ± 24	10 ± 1
Buckwheat	537 ± 27	28 ± 2
Mustard	469 ± 24	50 ± 9
Sicklepod	861 ± 73	60 ± 6
Velvetleaf	92 ± 7	50 ± 4
Jimson weed	114 ± 29	83 ± 16

[1] Values given for the antioxidants represent the mean ± SE of 6 plants of each species. Adapted from table 1 in ref. 32.

leaves of chamber-grown (38.2 ± 2.4 mg/100 g dry weight) compared to wild purslane (43.5 ± 3.0 mg/100 g dry weight), but these levels were lower than those present in spinach (63.5 ± 5.7 mg/100 g dry weight) (table 10).

Glutathione. The protective role of glutathione as an antioxidant and detoxifying agent has been demonstrated in various clinical studies. It is a ubiquitous compound that is synthesized rapidly in the liver, kidney, and other tissues, including the gastrointestinal tract. In animal cells, glutathione acts as a substrate for glutathione peroxidase, which reduces lipid peroxides that are formed from polyunsaturated FA (PUFA) in the diet, and as a substrate for glutathione-S transferase, which conjugates electrophilic compounds. Recent studies show that glutathione obtained from the diet is directly absorbed by the gastrointestinal tract and thus dietary glutathione can readily increase the antioxidant status in humans [35]. Dietary glutathione, in addition to levels supplied by the bile, may be used by the small intestine to decrease the absorption of peroxides. These results indicate that in the intact animal, lumenal glutathione is available for use by the intestinal epithelium to metabolize peroxides and other reactive species and to prevent their transport to other tissue.

Dietary glutathione occurs in highest amounts in fresh meats, in moderate amounts in some fruits and vegetables, whereas it is absent or found only in small

Table 12. Potential health effects of dietary glutathione in humans

---

Glutathione may protect cells from carcinogenic processes through a number of mechanisms:
(1)   by functioning as an antioxidant [35, 39]
(2)   by binding with mutagenic chemical compounds [40, 41]
(3)   by directly or indirectly acting to maintain functional levels of other antioxidants such as vitamins C and E and β-carotene [41–43]
(4)   through its involvement in DNA synthesis and repair [44, 45]
(5)   by enhancing the immune response [46, 47]

Adapted from Jones et al. [36].

---

amounts in grains and dairy products [36]. Only fresh asparagus at 28.3 mg/100 g and fresh avocado at 27.7 mg/100 g were higher than purslane in glutathione content in a study carried out to determine the glutathione content of 98 food items, identified by the National Cancer Institute, to contribute 90% or more of calories, dietary fiber, and 18 major nutrients in the US diet [36–38].

The potential health effects of dietary intake of glutathione in humans are shown in table 12 [35, 39–47]. In a recent study by Flagg et al. [48], plasma glutathione concentrations varied widely in humans and were influenced by sex and age (increased with age in men, but decreased with age and were lower in women who used estrogen-containing contraceptives).

Glutathione is now known to be widely distributed in plant cells and is the major free thiol in many higher plants [49–52]. Considerable variations in levels of glutathione have been reported by different studies recording thiol levels in a variety of plant species. This may be partially due to the use of different analytical techniques, and because glutathione levels vary both diurnally [53, 54] and with developmental and environmental factors [55–57]. Taking into account these considerations, the levels of glutathione found in purslane, 14.81 ± 0.78 mg/100 g fresh weight, were in the range of those reported for other plant species but significantly higher than the level of 9.65 ± 0.62 mg/100 g fresh weight for spinach (table 13). Glutathione was present in significantly greater amounts in chamber-grown purslane relative to wild plants, which may have reflected a difference in the developmental stage of the plants analyzed, or in the environmental conditions experienced.

Table 13. Glutathione content of purslane and spinach leaves[1]

	GSH	GSSX	GSH/GSSX
Chamber-grown purslane	14.81 ± 0.78 (0.48)	2.20 ± 0.15 (0.031)	6.73
Wild purslane	11.90 ± 0.63 (0.39)	1.42 ± 0.12 (0.023)	8.38
Spinach	9.65 ± 0.62 (0.31)	2.39 ± 0.20 (0.039)	4.03

Data represent mean values (mg/100 g fresh weight) from four analyses each with three replicates per plant species/type. Figures in parentheses are values expressed as μmol/g fresh weight to allow comparison with data previously reported in the literature. GSH = Gluta-thione; GSSX = glutathione-linked disulfides.

[1] In Simopoulos et al. [2].

Profiles of Leaf ω3 FA and Antioxidants throughout Plant Development in Growth Chamber Conditions

As discussed above, common purslane has recently been recognized as a wild plant species containing high levels of ω3 FA and α-tocopherol relative to green leafy vegetables currently common in the European and American diet. In a previous study [2], we analyzed leaf lipid and antioxidant content of plants grown in an environmental growth chamber for 28 days. However, lipid composition can vary with plant growth, age and environmental conditions. If purslane is to be developed as a commercial crop, it therefore becomes important to optimize nutritional quality in terms of growth stage and time of harvest. The report by Omara-Alwala et al. [30] showed that the FA content in purslane varied with the age of the plant, with the C_{20} and C_{22} ω3 FA increasing with age. We were unable to detect the presence of these compounds in the plants we analyzed at a relatively early developmental stage [2], and it is clear that more detailed studies are necessary. To approach this we prepared a profile of leaf FA and antioxidants throughout development of wild purslane grown from seed in a controlled environmental chamber [18-hour photoperiod (200 μmol m^{-2} s^{-1} PPFD) at 24°C with a 17°C dark period; Norman and Simopoulos, unpubl. data]. The growth conditions were maintained as in our previous study [2].

The content of α-tocopherol in leaf tissue increased to a maximum level of about 19 mg/100 g fresh weight at 45 days from planting, and then declined (table 14). Ascorbic acid content reached a maximum level (about 37 mg/100 g fresh weight) at about 40–45 days, and also subsequently declined. The level of gluta-thione increased from 40–45–50 days of growth, and showed a later decline after about 60 days. These changes in antioxidants in more mature leaves were not

Table 14. Antioxidant content during leaf development of purslane

Age, days	α-Tocopherol	Ascorbic acid	β-Carotene	Glutathione
25	9.4±0.3	24.2±0.7	1.7±0.1	14.0±0.6
30	11.7±0.4	28.6±0.8	2.0±0.1	15.8±0.8
40	16.3±0.4	36.8±0.6	2.4±0.2	20.3±0.5
45	18.9±0.5	37.2±0.4	2.6±0.2	23.4±0.7
50	15.6±0.3	27.6±0.5	2.7±0.3	24.0±0.8
60	11.0±0.4	25.3±0.7	2.5±0.1	19.6±0.6

Data are expressed as mg/100 g fresh weight.

associated with the initiation of chloroplast senescence as assessed by a change in chlorophyll content [58] which showed no significant decline during the 60 day growth period (data not shown). The content of β-carotene remained stable between 30–60 days in contrast to declines in other antioxidants (table 14).

The maximum level of 18:3 in developing leaf tissue occurred after about 50 days of growth in the environmental conditions we maintained (table 15). Both the absolute (table 15) and the relative amount (table 16) of 18:3 increased during this time, suggesting that both the overall rate of lipid synthesis and FA desaturase activity increased. Each of these processes appeared to decline between 50 and 60 days of growth (table 15 and 16). An early increase in PUFA content in developing leaf tissue has been previously reported in other plant species and may be related to chloroplast development and galactolipid biosynthesis [59]. The relative amount of 18:2 increased between 50 and 60 days (table 16), suggesting that desaturation of 18:2 to 18:3 may become inhibited with maturation. Chloroplast 18:2 desaturase activity has been stabilized in vitro by the addition of catalase [60], suggesting that the enzyme may be inactivated by peroxidation. We can speculate that the decline in levels of 18:3 in the more mature purslane leaves may be related to the decline in certain antioxidants. It has been suggested in particular that the role of chloroplast α-tocopherol may be to protect lipids from oxidation [61]. A relatively high concentration of this compound has been found in chloroplast envelopes [61], which are the site of galactolipid synthesis [62], and 18:2 desaturation [63]. Loss of galactolipids and LNA has generally been associated with chloroplast senescence and appears to involve lipases and peroxidation reactions [64]. However, the changes in purslane lipids were relatively small compared to those observed in late senescence [64] and we found no associated loss of chlorophyll (data not shown).

Table 15. FA content during leaf development of purslane

Age days	18:1	18:2	18:3	20:0	24:0	Total content
25	0.361	0.671	2.512	0.049	0.021	5.000
30	0.340	0.864	3.043	0.073	0.025	5.406
40	0.254	0.993	3.843	0.090	0.045	6.283
45	0.282	0.970	4.464	0.122	0.057	7.165
50	0.370	0.763	4.928	0.127	0.063	7.004
60	0.367	1.098	4.092	0.118	0.081	6.752

Data are expressed as mg/g fresh weight.

Table 16. FA composition of total lipid content during plant development

Age, days	18:1	18:2	18:3	20:0	24:0
25	7.22	13.42	50.24	0.98	0.42
30	6.30	15.98	57.21	1.35	0.40
40	4.04	15.80	61.16	1.43	0.72
45	3.93	13.54	62.30	1.70	0.79
50	5.28	10.89	70.36	1.81	0.90
60	5.43	16.26	60.60	1.75	1.20

Data represent wt%.

The levels of 20:0 and 24:0 remained relatively low throughout leaf development (table 15 and 16), and in this study we were unable to detect the 20:5, 22:5 and 22:6 FA previously reported in purslane [30]. The long-chain saturated FA such as 20:0 and 20:4 are normal components of plant waxes.

Overall, the results of this study indicate that in the growing conditions we maintained with this variety of purslane, the nutritional quality of leaf tissue, in terms of antioxidants and LNA content, would be optimal between 45 and 50 days. Obviously, the rate of plant development and leaf composition will be very different under field conditions. The growth chamber studies were useful in that they indicated an association between PUFA and antioxidant content.

In consideration of dietary sources of antioxidants, we would like to emphasize that plant leaf tissue in some unique aspects offers a direct source of antioxidant compounds naturally maintained in an active plant chloroplast. Accumulated evidence indicates definite requirements for specific antioxidant systems in photosynthetically active chloroplasts. During the process of photosynthesis (in which light energy is absorbed and transformed into reducing equivalents), various free radicals and oxidative products are normally generated. Given that the thylakoid membranes of plant chloroplasts are characteristically enriched in PUFA essential to maintain membrane structure and function, which are potentially highly susceptible to lipid peroxidation by reactive oxygen species, chloroplasts must maintain highly efficient free-radical scavenging systems for protection of lipids as well as pigments and specific enzyme systems. In particular, the environment within an active chloroplast during photosynthesis favors the production of singlet oxygen. Linked with this is an active electron transport system with the potential for electron donation. Chloroplast antioxidants include glutathione and ascorbate, as well as α-tocopherol, and carotenoids, and associated enzyme systems. In general, both ascorbate and glutathione are present in plant chloroplasts at millimolar concentrations, and a high concentration of the lipid-soluble α-tocopherol has been found in specific regions of the plant chloroplast [65]. The available evidence from plant studies indicates that the following enzymes and antioxidants are critical in developing leaf tissue: ascorbate, α-tocopherol, reduced glutathione, carotenoid pigments, superoxide dismutase, ascorbate peroxidase, dehydroascorbate reductase, and glutathione reductase [65]. Additional research with plant cells indicates that the antioxidant action of α-tocopherol may include a synergism with ascorbate as well as a direct interaction with free-radical species [65]. Other antioxidants include certain flavonols and possibly polyamines. Some basic antioxidant reactions operative in chloroplast during photosynthesis are shown in figure 2. Superoxide is scavenged by superoxide dismutase enzymes to produce hydrogen peroxide. However, this is potentially deleterious since hydrogen peroxide is a potent inhibitor of several enzymes including superoxide dismutase and specific bisphosphatases of the Calvin cycle [65]. This process is achieved via the glutathione cycle (fig. 2), which includes enzymes ascorbate peroxidase, dehydroascorbate peroxidase, and glutathione reductase, and leads to the maintenance of reduced glutathione and ascorbate.

Clearly from our studies, with the leaf tissue of *P. oleracea* and possibly also the seeds (from previous studies), we can emphasize that plant tissues should be reevaluated as a potential dietary source of both essential FA and antioxidants. Purslane is certainly outstanding in this respect.

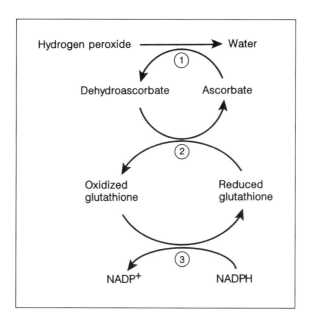

Fig. 2. The ascorbate-glutathione cycle in plant chloroplasts. Enzymes involved are: (1) ascorbate peroxidase; (2) dehydroascorbate reductase, (3) glutathione reductase.

Agricultural Potential of Purslane

Earlier studies by Kabulov and Tashbekov [66] emphasized the nutritional benefits of purslane. Although these workers did not determine FA composition, they carried out agricultural studies for the purpose of introducing it into cultivation. They referred to purslane as being rich in different salts and especially in proteins and carbohydrates. The maximum concentration of crude protein (23.7%) and carbohydrates (13.8%) was noted in plants during the period of seed maturity, of β-carotene (16.8 mg%) and ascorbic acid (231 mg%) at the beginning of vegetation and of nicotinic acid (50 mg%) and α-tocopherol (6 mg%) in the flowering phase.

Purslane is highly efficient in water use and grows in arid lands (fig. 3). Due to the increasing amount of dry lands in the world, Miller et al. [29] studied the arid-land agricultural potential of three specific C_4 species – *Amaranthus hypochondriacus* (prince's feather), *Amaranthus retroflexus* (redroot pigweed), and *P. oleracea* (common purslane) – and contrasted them with two C_3 species – *Beta vulgaris* variety *cicla* (Swiss chard) and *Chenopodium album* (lamb's quarters) – under different conditions of temperature, water availability and soil salinity. Purslane seed had the highest FA content of all five species (21.5%) and the high-

Fig. 3. Wild purslane plant flourishing in an arid environment. Courtesy of Don Pusateri-Nutrilite.

est protein content (19.9%) which is relatively high (>18% protein) when compared to cereals such as wheat (16.1%), barley (9.2%), and corn (10.3%) [29].

The FA content of leaves was slightly higher and the protein was significantly higher when less water was available.

Table 4 shows the amino acid composition of purslane seeds in comparison to egg and soybean. Purslane has the best balance and highest concentration of essential amino acids and a high chemical score in comparison to the other C_4 and C_3 species studied by Miller et al. [29].

The leaf content of individual amino acids was not affected by the amount of water available, although the percent of crude protein was highest under the lower-watering treatment. Miller concluded:

'The germination and growth characteristics of *Portulaca oleracea* indicate that this plant species is particularly well suited for growth in arid environments. The seeds of *P. oleracea* have high concentrations of both fat and protein and are of sufficient quality for use as a ruminant feed or as a grain lysine supplement. However, the high fiber content of the seeds may limit their use to ruminant animals. Further studies need to be done on this promising species.'

Development of purslane as a commercial item of conventional agriculture would appear feasible (fig. 4, 5), if demand for direct human consumption or for

Fig. 4. Harvesting of cultivated purslane. Courtesy of Don Pusateri-Nutrilite.

Fig. 5. Cultivated purslane. Courtesy of Don Pusateri-Nutrilite.

livestock feeding warrant. Research has been directed at its potential as a crop for arid lands [29]. It would appear to offer much promise for this purpose, and the threat of global warming could lend some urgency to this line of investigation. Purslane, along with crop plants that grow rapidly in bright sunlight like corn and sugarcane, has C_4 metabolism, the initial product of CO_2 fixation being a 4-carbon molecule. Such plants are able to fix CO_2 and carry on effective photosynthesis under conditions of dryness, high temperature and intense lighting. C_3 plants under the same conditions engage in an unproductive metabolism called photorespiration. C_4 plants are also characterized by high growth rates and water use efficiency. In the case of purslane this efficiency is so high that research was undertaken to detect possible presence of the process of crassulacean acid metabolism, which would further increase water efficiency by fixation and storage of CO_2 in acid form at night. Many similarities to the crassulacean type were found, allowing the possibility that favorable water status earlier in the season would permit rapid C_4 photosynthesis, while water stress later would induce and be survived through the crassulacean type [67, 68].

In a study on the chemical composition of purslane, Vengris et al. [69] reported that common purslane had the highest content of potassium of all investigated common weeds in Massachusetts (USA).

Medicinal Uses of Purslane

Purslane was used for its medicinal properties for thousands of years by the ancient Greeks, Persians, Indians and throughout Africa and in many countries around the world (table 17). Known since the time of Hippocrates, purslane was used by Theophrastus and Dioscorides for its diuretic, anthelminthic, antiscorbutic and cathartic properties. As can be seen from table 18 [28], some of the medicinal uses attributed to purslane, such as its anti-inflammatory and diuretic properties and its use to improve arthritis, headaches, shortness of breath and as a cardiac tonic, are similar to the effects attributed to ω3 FA [3–8] and to antioxidant vitamins and glutathione.

As can be seen in table 17, the medicinal uses of purslane are extensive. For this reason, purslane has been investigated for a number of pharmacologically active substances.

High concentration of noradrenaline in purslane was reported by Feng et al. [70]. Both by paper chromatography and biological estimation, it was found that a crude extract contained noradrenaline approximately equivalent to 2.5 mg/g of fresh weight. The authors commented: 'It is of interest to note that the concentration of noradrenaline in this plant might be greater than that extractable from suprarenal glands of mammals' [70].

Table 17. Medicinal uses of purslane in various countries

Country	Uses
Columbia	Emollient on tumors, calluses
Philippines	Heal burns, skin diseases
China	Emollient, leaves used as poultice for tumors, bad wounds and ulcers, whereas the seeds are considered a diuretic
Gold Coast	Leaves ground, mixed with oil, used on boils to bring them to a head
West Tropical Africa	Leaves used as a poultice on boils and burns, and as heart tonic and diuretic
Nigeria	Leaves applied to swellings
Siberia	Eaten as a regular food, used as a gastric sedative herb, for prickly heat, applied to forehead and temple to relieve heat and pain, applied to eyes to remove inflammation
West Indies, Cochin-China	The Tamil practitioners use seeds for stomach problems, to provoke menses; also as an emollient and diuretic. The bruised fresh leaves are used externally for erysipelas
Guadalupe, Corre, Lejanne	The whole plant is used as a tonic and febrifuge
Jamaica	The plant is given as a 'cooling medicine' for fevers
North America	The whole plant is used as a cooling diuretic, whereas the seeds are considered to be anthelminthic
Punjab	Seeds are used as a vermifuge
Punjab and Cashmere	Seeds are used by the hakims in inflammation of the stomach and intestinal ulcerations

Hegnauer [71] in 1969 reported that aqueous extracts of purslane contain dopa, dopamine, catecholamines and noradrenaline. Dopa and dopamine are known to be common components of plants. They are found in bananas, plantains and potatoes. However, the amount of noradrenaline found in purslane is much greater than in any of these plants. In 1986, Okwuasaba et al. [72] carried out pharmacological studies on in vitro skeletal muscle preparations and showed that aqueous extracts of purslane possessed muscle-relaxant properties most likely due to interference with mobilization of calcium with consequent impairment of the excitation-contraction coupling mechanism. Parry et al. [73] reported that aqueous extracts applied topically onto the skin in a young female patient decreased muscle spasticity due to an incomplete T_6 lesion of the spinal cord as a result of an injury.

Aqueous extracts of purslane produced skeletal muscle relaxation in rats following oral or intraperitoneal administration. In these experiments, the aqueous

Table 18. Purslane: medicinal uses and attributes[1]

Burns and trauma	Cardiac tonic
Headaches	(improves heart contractions)
Stomach ailments	Arthritis
Intestinal ailments	Cathartic
Liver ailments	Anthelminthic
Cough	Antiscorbutic attributes
Shortness of breath	Anti-inflammatory
	Diuretic

[1] In Simopoulos [28].

extract given intraperitoneally (200–1,000 mg/kg) proved more effective skeletal muscle relaxant than chlordiazepoxide (20 mg/kg, i.p.), diazepam (40 mg/kg, i.p.) and dantrolene sodium (30 mg/kg, orally) [74].

Additional studies by Parry et al. [75] in rats, rabbits and guinea pigs suggested that the aqueous extract from purslane leaves and stems acts in part on postsynaptic α-adrenoreceptors and by interference with transmembrane calcium influx, most likely due to ω3 FA content. Parry et al. [75, 76] attributed the skeletal muscle relaxant properties, possibly due 'to an interference with Ca^{++} mobilization and with subsequent impairment of the excitation-contraction coupling mechanisms'.

Alcohol extracts of purslane seeds administered subcutaneously in rodents effectively impaired spermatogenesis after 30 dose treatments [77], suggesting that purslane could be used as a male contraceptive.

Contribution of Purslane to Food Technology: Pectin

Pectins are a group of heterogeneous polysaccharides with a high molecular weight [78]. The Joint FAO/WHO Expert Committee on Food Additives has evaluated pectin and cleared it toxicologically. The Committee considered it unnecessary to establish an acceptable daily intake, since pectin is recognized as a valuable and harmless food additive. Therefore, it is included on the list of permitted additives in standardized foods, if a technological need can be proven. Wenzel et al. [79] extracted a polysaccharide complex in yields up to 25 g% (dry weight). The physicochemical properties of the clear and viscous mucilage render it appropriate for industrial uses such as food extenders and viscosifier. Further fractionation of the crude extract by anion exchange chromatography separated the crude extract into a neutral arabinogalactan and polydisperse pectin-like poly-saccharides.

Pectins occur in the intercellular regions and cell walls of most fruits and vegetables and are found in abundance in oranges, grapefruit, lemons and limes [80]. Citrus pectin has been extensively studied for its hypolipidemic properties. Pectin lowers LDL cholesterol while only slightly lowering HDL in rodents, and miniature swine by 30% [81–83]. In the latter, pectin ingestion reduced athero-sclerosis after prolonged hypercholesterolemia. Most recently, studies in minia-ture pigs showed that ingestion of 3% pectin reduced platelet aggregation before any change in plasma cholesterol occurred. Therefore reduction in platelet aggre-gation may contribute to the anti-atherogenic effects of pectin [84].

Pectin is difficult to incorporate into foods and beverages due to its thicken-ing properties. However, Cerda et al. [85] developed an egg white-pectin complex in order to incorporate pectin into a variety of food products and tested this prod-uct in guinea pigs to evaluate this animal as a model for cholesterol research. In this model, the egg white-pectin incorporated into foods produced a significantly lower plasma cholesterol than observed in the controls [85]. Studies in humans with hypercholesterolemia indicate that grapefruit pectin in the form of capsules providing 15 g of pectin per day given as 5 g per meal for three meals, led to a 7.6% decrease in plasma cholesterol level at the end of 4 weeks. This study showed that daily dietary supplement of 15 g of pectin significantly lowered plas-ma cholesterol and improved the ratio of LDL:HDL in hypercholesterolemic patients who were unable or unwilling to follow a low-risk (low cholesterol) diet. It can be concluded that grapefruit pectin and other food sources rich in pectin are useful adjuvants to a fat-restricted diet in hypercholesterolemic patients [86]. The high pectin content in purslane leaves, in addition to the ω3 FA and antioxidant content, is another reason for the medicinal uses of purslane as a cardiac 'tonic' by so many groups around the world since prehistoric times.

Conclusions

Wild plants have contributed to the diet of both human beings and animals since their first appearance on planet earth. Humans ate a variety of wild plants, whereas today the diet of developed societies is limited to a few cultivated vegeta-bles. Epidemiologic studies show that populations whose diets are rich in fruits and vegetables have lower death rates from cancer, coronary heart disease, dia-betes and gastrointestinal disorders. Because plants are high in fiber content, fiber was considered to be the major factor contributing to differences between the developed and developing countries in terms of the prevalence of chronic diseases. The contributions of green leafy vegetables in terms of ω3 FA and antioxidants were not considered.

We were interested in the ω3 FA and antioxidant vitamin content of purslane because of its reported medicinal applications for acute and chronic conditions, and as a food. Our studies, and those of others, clearly show that purslane is an excellent source of LNA, vitamin E, vitamin C, β-carotene, glutathione, potassium and pectin, all of which may contribute to the medicinal uses of purslane as a cardiac tonic, diuretic, against infections of the skin, and disorders of the gastrointestinal tract. The high content of dopa and the pharmacologic studies reported suggest that additional research is needed on the composition of purslane for the identification of substances responsible for decreasing spermatogenesis and its use as a 'natural' contraceptive agent in males.

Because purslane has a C_4 metabolism and can be grown in arid lands, the potential of purslane as an agricultural commodity and food for humans has led some to consider purslane as a 'power food of the future' [87]. Based on our research, purslane and other wild plants ought to once again be incorporated into the diet of human beings, both in developed and developing countries. Among 13,000 known food plants, fewer than 20 are currently providing most of our food needs, yet many of the underutilized plants, such as purslane, offer better nourishment than the major crops and can grow in poor soil, in droughts, and in arid climates.

References

1   Simopoulos AP, Salem N Jr: Purslane: A terrestrial source of omega-3 fatty acids. N Engl J Med 1986;315:833.
2   Simopoulos AP, Norman HA, Gillaspy JE, Duke JA: Common purslane: A source of omega-3 fatty acids and antioxidants. J Am Coll Nutr 1992;11:374–382.
3   Simopoulos AP: Omega-3 fatty acids in health and disease and in growth and development. Am J Clin Nutr 1991;54:438–463.
4   Simopoulos AP; Kifer RR, Martin RE (eds): Health Effects of Polyunsaturated Fatty Acids in Seafoods. Orlando, Academic Press, 1985.
5   Galli C, Simopoulos AP (eds): Dietary ω3 and ω6 Fatty Acids. Biological Effects and Nutritional Essentiality, Series A: Life Sciences. New York, Plenum Press, 1989, vol 171.
6   Simopoulos AP, Kifer RR, Martin RE, Barlow SM: Health Effects of ω3 Polyunsaturated Fatty Acids in Seafoods. World Rev Nutr Diet. Basel, Karger, 1991, vol 66.
7   Effects of Fish Oils and Polyunsaturated Omega-3 Fatty Acids in Health and Disease. DHHS Publication. Special Bibliography No. 1992-A. Bethesda, Md., Public Health Service, National Institutes of Health, National Library of Medicine.
8   Galli C, Simopoulos AP, Tremoli E (eds): Fatty Acids and Lipids from Cell Biology to Human Disease. World Rev Nutr Diet. Basel, Karger, 1994, vol 75.
9   Diet and Health: Implications for Reducing Chronic Disease Risk. National Research Council. Washington, National Academy Press, 1989.
10 National Research Council: Diet, Nutrition, and Cancer. Report of the Committee on Diet, Nutrition, and Cancer. Assembly of Life Sciences. Washington, National Academy Press, 1982.

11 Slater TF, Block G (eds): Antioxidant vitamins and beta-carotene in disease prevention. Am J Clin Nutr 1991;53:189s–396s.

12 Rifici VA, Khachadurian AK: Dietary supplementation with vitamin C and E inhibits in vitro oxidation of lipoproteins. J Am Coll Nutr 1993;12:631–637.

13 Eaton SB, Konner M: Paleolithic nutrition. A consideration of its nature and current implications. N Engl J Med 1985;312:283–289.

14 Coquillat M: Sur les plantes les plus communes à la surface du globe. Bull Mens Soc Linn Lyon 1951;165–170.

15 Holm LG, Plucknett DL, Pancho JV, Herberger JP: The World's Worst Weeds. Distribution and Biology. Honolulu, University of Hawaii Press, 1977, vol 7, pp 78–83.

16 Byrne R, McAndrews JH: Pre-Columbian purslane (*Portulaca oleracea*) in the new world. Nature 1975;253:726–727.

17 Chapman J, Stewart RB, Yarnell RA: Archaelogical evidence for precolumbian introduction of *Portulaca oleracea* and *Mollugo verticillata* into eastern North American. Econ Bot 1974;28:411–412.

18 Kaplan L: Ethnobotany of the Apple Creek archaeological site, southern Illinois. Am J Bot 1973;60(suppl 4):39.

19 Watson PJ: The prehistory of Salts Cave, Kentucky. Rep Invest Springfield, Ill State Mus 1969;16.

20 Watson PJ (ed): Archaeology of the Mammoth Cave area. New York, Academic Press, 1974.

21 Walker WM: The Troyville Mounds, Catahoula Parish, Louisiana. Bur Am Ethnol Bull 1936;113.

22 Danin A, Baker I, Baker HG: Cytogeography and taxonomy of the *Portulaca oleracea* L. polyploid complex. Isr J Bot 1978;27:177–211.

23 Simpson JA, Weiner ESC: The Oxford English Dictionary, ed 2. Oxford, Clarendon Press, 1989, vol 12, p 886.

24 Clarcke CB: Edible and Useful Plants of California. Berkeley, University of California Press, 1977, pp 208–209.

25 Halpin AM: Unusual Vegetables. Emmaus, Rodale Press, 1978, pp 323–325.

26 Spencer ER: All about Weeds. New York, Dover Publishing, 1957, pp 119–121.

27 American Horticulturist: Strange relatives: The purslane family. June 1985, pp 5–8.

28 Simopoulos AP: Terrestrial sources of omega-3 fatty acids: Purslane; in Quebedeaux B, Bliss F (eds): Horticulture and Human Health: Contributions of Fruits and Vegetables. Englewood Cliffs, Prentice-Hall, 1987, pp 93–107.

29 Miller TE, Wing JS, Huete AR: The agricultural potential of selected C4 plants in arid environments. J Arid Environ 1984;7:275–286.

30 Omara-Alwala TR, Mebrahtu T, Prior DE, Ezekwe MO: ω-Three fatty acids in purslane (*Portulaca oleracea*) tissues. J Am Oil Chem Soc 1991;68:198–199.

31 Harwood JL: Plant acyl lipids: Structure, distribution and analysis; in Stumpf PK (ed): The Biochemistry of Plants. Lipids: Structure and Function, New York, Academic, Press, 1980, vol 4, pp 1–55.

32 Finckh BF, Kunert KJ: Vitamin C and E: An antioxidative system against herbicide-induced lipid peroxidation in higher plants. J Agric Food Chem 1985;33:574–577.

33 Spurgeon SL, Porter JW: Carotenoids; in Stumpf PK (ed): The Biochemistry of Plants, Lipids: Structure and Function. New York, Academic Press, 1980, vol 40, pp 419–483.

34 Goodwin TW: Distribution of carotenoids; in Goodwin TW (ed): Chemistry and Biochemistry of Plant Pigments. New York, Academic Press, 1976, vol 1, pp 127–142.

35 Jones DP, Hagen TM, Weber R, Wierzbicka GT, Bonkovsky HL: Oral administration of glutathione (GSH) increases plasma GSH concentrations in humans (abstract). FASEB J 1989;3:A1250.

36 Jones DP, Coates RJ, Flagg EW, et al: Glutathione in Foods listed in the National Cancer Institute's Health Habits and History Food Frequency Questionnaire. Nutr Cancer 1992;17:57–75.

37 Block G, Dresser CM, Hartman AM, Carroll MD: Nutrient sources in the American diet: Quantitative data from the NHANES II Survey. I. Vitamins and Minerals. Am J Epidemiol 1985;122:13–26.

38   Block G, Dresser CM, Hartman AM, Carroll MD: Nutrient sources in the American diet: Quantitative data from the NHANES II Survey. II. Macronutrients and Fats. Am J Epidemiol 1985;122: 27–40.

39   Mannervik B, Carlberg I, Larson K: Glutathione: General review of mechanism of action; in Dolphin D, Avramovic O, Pulson R (eds): Glutathione. Chemical, Biochemical and Medical Aspects. New York, Wiley, 1989, pt A, pp 475–516.

40   Wattenberg LW: Perspectives in cancer research. Chemoprevention of cancer. Cancer Res 1985;45: 1–8.

41   Frei B, England L, Ames BN: Ascorbate is an outstanding antioxidant in human blood plasma. Proc Natl Acad Sci USA 1989;86:6377–6381.

42   Bendich A: Antioxidant micronutrients in immune responses; in Bendich A, Chandra RK (eds): Micronutrients and Immune Functions. New York, New York Academy of Science, 1985, vol 587, pp 169–180.

43   Frei B, Stocker R, Ames BN: Antioxidant defenses and lipid peroxidation in human blood. Proc Natl Acad Sci USA 1988;85:9748–9752.

44   Oleinick NL, Xue L, Friedman LR, Donahue LL, Biaglow JE: Inhibition of radiation-induced DNA-protein cross-link repair by glutathione depletion with L-buthionine sulfoximine. NCI Monogr 1988;6:225–229.

45   Fuchs JA: Glutaredoxin; in Dolphin D, Avramovic O, Poulson R (eds): Glutathione. Biochemical and Biochemical and Medical Aspects. New York, Wiley, 1989, pt B, pp 551–570.

46   Furukawa T, Meydani SN, Blumberg JB: Reversal of age-associated decline in immune responsiveness by dietary glutathione supplementation in mice. Mech Ageing Dev 1987;38:107–117.

47   Buhl R, Holroyd KJ, Mastrangeli A, Cantin AM, Jaffe HA, et al: Systemic glutathione deficiency in symptom-free HIV-seropositive individuals. Lancet 1989;ii:1294–1298.

48   Flagg EW, Coates RJ, Jones DP, et al: Plasma total glutathione in humans and its association with demographic and health-related factors. Br J Nutr, in press.

49   McCay PB: Vitamin E: Interactions with free radicals and ascorbate. Annu Rev Nutr 1985;5:323–340.

50   Renneberg H: Glutathione metabolism and possible biological roles in the higher plant. Phytochemistry 1982;21:2771–2781.

51   Renneberg H: Aspects of glutathione function and metabolism in plants; in Von Wettstein D, Chua NH (eds): Plant Molecular Biology. New York, Plenum Press, 1987, pp 279–292.

52   Hatzios KK, Bormann JF (eds): Proceedings 1989 Annual Symposium South ASPP: Glutathione Synthesis and Function in Higher Plants. Physiol Plant 1989;77:447–471.

53   Koike S, Patterson BD: Diurnal variation of glutathione levels in tomato seedlings. Hort Sci 1988; 23:713–714.

54   Schupp R, Rennenberg H: Diurnal changes in the glutathione content of spruce needles (*Picea abies* L.) Plant Sci 1988;57:113–117.

55   Earnshaw BA, Johnson MA: Control of wild carrot somatic embryo development by antioxidants. Plant Physiol 1987;85:273–276.

56   De Kok LJ, De Kan PJL, Tanczos OG, Kuiper PJC: Sulphate-induced accumulation of glutathione and frost-tolerance of spinach leaf tissue. Physiol Plant 1981;53:435–438.

57   Wise RR, Naylor AW: Chilling-enhanced photooxidation. The peroxidative destruction of lipids during chilling injury to photosynthesis and ultrastructure. Plant Physiol 1987;83:272–277.

58   Gepstein S: Photosynthesis; in Nooden LD, Leopold AC (eds): Senescence and Aging in Plants. San Diego, Academic Press, 1988, pp 85–104.

59   Harwood JL: Fatty acid metabolism. Annu Rev Plant Physiol Plant Mol Biol 1988;39:101–138.

60   Norman HA, Pillai P, St. John JB: In vitro desaturation of monogalactosyldiacylglycerol and phosphatidylcholine molecular species by chloroplast homogenates. Phytochemistry 1991;30:2217–2222.

61   Douce R, Block MA, Dorne A-J, Joyard J: The plastid envelope membranes: Their structure, composition, and role in chloroplast biogenesis; in Roodyn DB (ed): Subcellular Biochemistry. New York, Plenum Press, 1984, vol 10, pp 1–84.

62   Heemskerk JWM, Wintermans JRGM: Role of the chloroplast in the leaf acyl lipid synthesis. Physiol Plant 1987;70:558–568.

63   Schmidt H, Heinz E: Desaturation of oleoyl groups in envelope membranes from spinach chloroplasts. Proc Natl Acad Sci USA 1990;87:9477–9480.

64   Gepstein S: Photosynthesis; in Nooden CLD, Leopold AC (eds): Senescence and Aging in Plants. San Diego, Academic Press, 1988, pp 85–104.

65   Kunert KJ, Dodge AD: Herbicide radical damage & antioxidative systems; in Boger P, Sandmann G (eds): Target Sites of Herbicide Action. Boca Raton, CRC Press, 1989, pp 15–62.

66   Kabulov DT, Tashbekov I: Purslane. Kartofel' i Ovoschi 1979;8:45–46.

67   Koch K, Kennedy RA: Characteristics of crassulacean acid metabolism in the succulent C4 dicot, *Portulaca oleracea* L. Plant Physiol 1980;65:193–7.

68   Welkie GW, Caldwell M: Leaf anatomy of species in some dicotyledon families as related to the C3 and C4 pathways of carbon fixation. Can J Bot 1970;48:2135–2146.

69   Vengris J, Drake M, Colby WG, Bart J: Chemical composition of weeds and accompanying crop plants. Agron J 1953;45:213–218.

70   Feng PC, Haynes LJ, Magnus KE: High concentration of (–)-noradrenaline in *Portulaca oleracea* L. Nature 1961;191:1108.

71   Hegnauer R: Chemotaxonomie der Pflanzen. 1969;5:383.

72   Okwuasaba F, Ejike C, Parry O: Skeletal muscle relaxant properties of the aqueous extract of *Portulaca oleracea*. J Ethnopharmacol 1986;17:139–160.

73   Parry O, Okwuasaba F, Ejike C: Preliminary clinical investigation into the muscle relaxant actions of an aqueous extract of *Portulaca oleracea* applied topically. J Ethnopharmacol 1987;21:99–106.

74   Parry O, Okwuasaba FK, Ejike: Skeletal muscle relaxant action of an aqueous extract of *Portulaca oleracea* in the rat. J Ethnopharmacol 1987;19:247–253.

75   Parry O, Okwuasaba, Ejike C: Effect of an aqueous extract of *Portulaca oleracea* leaves on smooth muscle and rat blood pressure. J Ethnopharmacol 1988;22:33–44.

76   Parry O, Okwuasaba F, Ejike C: Skeltal muscle relaxant actions of an aqueous extract of *Portulaca oleracea* in the rat. J Ethnoparmacol 1987;19:247–253.

77   Verma OP, Kumar S, Chatterjee SN: Antifertility effects of common edible *Portulaca oleracea* on the reproductive organs of male albino mice. Indian J Med Res 1982;75:301–310.

78   Nelson D, Smith JB, Wiles R: Commercially important pectinic substances; in Graham HD (ed): Food Colloids. Westport, AVI Publishing, 1977, pp 418–422.

79   Wenzel GE, Fontana JD, Correa JBC: The viscous mucilage from the weed *Portulaca oleracea*, L. Appl Biochem Biotechnol 1990;24/25:341–353.

80   Rouse A: Pectin: Distribution, significance; in Nagy S, Shaw P, Veldhuis M (eds): Citrus Science and Technology. Westport, AVI Publishing, 1977, vol 1, pp 110–113.

81   Baig MM, Cerda JJ: Pectin: Its interaction with serum lipoproteins. Am J Clin Nutr 1981;34:50–53.

82   Baekey PA, Robbins FL, Burgin CW, Cerda JJ: The effect of citrus pectin on the development of atherosclerosis in miniature swine. Clin Res 1986;34:388.

83   Baekey AP, Cerda JJ; Burgin CW, Robbins FL, Baumgartner TG: Effect of grapefruit pectin on plasma cholesterol and development of atherosclerosis in miniature swine. JPEN 1987;11:85.

84   Benjamin MA, Von der Porten AE, Cerda JJ, Mehta JL: Pectin administration decreases platelet aggregation in cholesterol-fed pigs without affecting cholesterol levels (abstract). 41st Annual Scientific Session of the American College of Cardiology, Dallas, April 1992.

85   Cerda JJ; Robbins FL, Burgin CW, Vathada S, Sullivan MP: Effectiveness of a grapefruit pectin product in lowering plasma cholesterol in guinea pigs. Clin Res 1991;39:9–801A.

86   Cerda JJ, Robbins FL, Burgin CW, Baumgartner TG, Rice RW: The effects of grapefruit pectin on patients at risk for coronary heart disease without altering diet or lifestyle. Clin Cardiol 1988;11:589–594.

87   Levey GA: The new power foods. Parade Magazine. The Washington Post. Sunday, November 14, 1993, p 5.

Artemis P. Simopoulos, MD, The Center for Genetics, Nutrition and Health,
2001 S Street, N.W., Suite 530, Washington, DC 20009 (USA)

Simopoulos AP (ed): Plants in Human Nutrition.
World Rev Nutr Diet. Basel, Karger, 1995, vol 77, pp 75–88

....

Sweet Lupins in Human Nutrition

Ricardo Uauy, Vivien Gattas, Enrique Yañez

Instituto de Nutrición y Tecnología de los Alimentos, Universidad de Chile,
Santiago, Chile

Contents

Introduction 75
Nutritional Value of Lupin as a Food 77
Nutritional Value of Lupin Proteins for Humans 80
Significance of Sweet Lupins as Plantfoods for Humans 82
Acknowledgments 86
References 86

Introduction

Grain legumes are important in human protein nutrition. Over 600 genera and 13,000 species of legumes have been described, yet only a few hundred have a place in the human diet, and most cultures traditionally consume less than 10 legume species. Lupins are an ancient leguminous plant used as a food by people surrounding the Mediterranean sea and those living in the Andean highlands. Lupins have been cultivated for the last six millennia in the New and for over 3,000 years in the Old world. The Egyptians, Greeks and Romans used *Lupinus albus* as a grain and as a soil enricher; since early times the plant was noted for its ability to grow in poor soils and in adverse climatic conditions. The grains were used as human food during pre-Columbian times. Native South Americans consumed lupins after soaking and cooking; this helped remove the bitter taste caused by the alkaloid. The Andean highland civilizations cultivated *L. mutabilis* extensively and toasted or cooked the grains to make them edible [1–3].

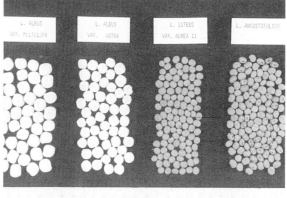

Fig. 1. Seeds of lupin species commonly cultivated in Chile, flowers correspond to *Lupinus albus* cv. Multolupa.

In modern times, King Frederick II of Prussia introduced *L. albus* to Germany during the 18th century. Results were poorer than expected, yet the cultivation of lupins in the sandy Baltic coastal plains prospered. It was used predominantly as an animal feed and as a soil enricher given its nitrogen-fixing properties. The appearance of lupin-associated toxicity in sheep led to diminished interest in their cultivation up to World War I. Renewed attention given to lupins during the war in Germany was justified based on the urgent need for alternate protein sources. Given the fact that the British navy established a blockade of sea lanes to Germany, lupins were one of the few legumes that could be cultivated in that region of the globe. Scientists at the Kaiser Wilhelm Research Institute were successful in removing the alkaloids and later selected lupins free of alkaloids, the so-called sweet lupins. Von Sengbusch was able to lower the alkaloid content of *L. albus*, *L.*

luteus and *L. angustifolius* from the traditional 1–3% down to less than 0.02%, thus obtaining the first true sweet lupins. Figure 1 illustrates the flower and seeds of four lupin cultivars [2, 4, 5].

Present geographic distribution of lupin species has followed migratory patterns. *L. albus*, *L. pilosus*, *L. varius* and *L. digitatus* followed the European migration to the Americas. *L. consentinii* and *L. angustifolius* reached Australia early in this century. Cultivation was successfully established in the poor sandy soils of the Western coast of Australia. They are now widely used for feeding sheep and other animals. Gladstones has reported that over 150 lupin species are cultivated globally. Australia, Poland, Germany, Russia and other members of the Commonwealth of Independent States account for over 90% of the world lupin production [5, 6].

Over the past four decades, the advent of new genetically selected varieties has led to seeds with greater protein and oil content, thinner testas and nonshattering pods. *L. luteus* is most prevalent in the Baltic coastal plains, *L. pilosus* and other large-seed species are commonly grown in the Mediterranean region. In the Andean region, researchers have improved the native *L. mutabilis* and *L. albus* by genetic selection. In collaboration with German scientists, they have generated several low-alkaloid cultivars with improved resistance to adverse climatic conditions. Hybrids have also been successfully obtained; best results have emerged from crossing *L. angustifolius* and *L. limifolius* [2, 5–8].

Nutritional Value of Lupin as a Food

The initial step in defining the nutritional value of a food is to address its chemical composition. Lupin seeds have a relatively stable composition, although cultivation conditions can modify the composition. Larger and fuller seeds have more protein and less crude fiber. During industrial processing, dehulling reduces the fiber while increasing protein content. The chemical composition of various species and cultivars of intact and dehulled lupin seeds is given in table 1. The testa's relative contribution to total dry weight may vary from 12% for *L. mutabilis*, 20% for *L. angustifolius* to 25% for *L. luteus* cultivar Aurea and >30% in *L. consentii*. The lower the testa's weight, the higher the relative oil and protein content. This is of relevance in animal nutrition where whole seeds may be used in the preparation of feeds. The crude fiber content for whole seeds varies from 11 to 17%, over 80% of this is in the hull. The seed after losing the hull has less than 2% crude fiber [9–13].

The protein content of whole lupin seeds is as high or higher than that of soya beans. *L. albus* contains close to 35% protein on a dry weight basis while *L. luteus* and *L. mutabilis* may have up to 44% protein. The protein content increases after

Table 1. Proximate composition of sweet lupins cultivated in Chile (g/100 g dry matter)

Lupinus cultivars	Moisture	Ash	Protein (N × 6.25)	Ether extract	Crude fiber	N-free extract
Albus						
Multolupa						
Whole seed	9.0	3.6	34.7	12.8	11.3	37.6
Kernel	8.0	3.8	39.6	13.5	2.1	41.0
Hull	7.5	2.8	3.0	0.6	54.7	38.9
Albus						
Astra						
Whole seed	8.5	3.8	35.3	10.7	13.1	37.1
Kernel	8.1	4.1	42.4	12.9	1.6	39.0
Hull	8.1	2.9	3.0	0.7	56.4	37.0
Luteus						
Aurea						
Whole seed	9.9	3.9	44.4	5.4	16.8	29.5
Kernel	8.2	4.5	56.6	6.8	2.0	30.1
Hull	7.6	2.1	3.6	0.6	59.0	34.7
Angustifolius						
Whole seed	10.2	3.2	34.2	5.6	13.1	43.9
Kernel	9.7	3.3	42.2	7.0	2.0	45.5
Hull	8.5	3.0	2.5	0.8	55.4	38.3

Adapted from reference 14.

dehulling. These values are nearly double those of legumes commonly used as human foods. The protein quality of a food can be predicted based on its amino acid composition and the protein digestibility. Thus, it is important to consider not only the total amount of protein but also the amino acid composition. Lupin protein, as is true for all legumes, has a low sulfur amino acid content. Table 2 summarizes the amino acid composition of protein from several lupin species and cultivars. The first limiting amino acids are the sulfur-containing acids, methionine and cystine; they are also relatively low in valine and tryptophan while high in lysine. Lupin protein is high in glutamic, aspartic acid and arginine. The pattern is typical for a legume protein and indicates possible complementarity with cereal grains in terms of lysine and methionine [9–17].

The oil content of lupin is much lower than that of soya. *L. albus* has approximately 11% lipids while other species have less than 6%, however, *L. mutabilis*

Table 2. Amino acid composition of sweet lupins (g/100 g of protein)

Amino acid	*Albus* astra	*Albus* multolupa	*Luteus* aurea	*Angusti-folius*	FAO/WHO[1] reference (1985)
Essential					
Isoleucine	3.1	4.1	3.1	3.3	5.4
Leucine	6.7	7.4	7.3	6.7	8.6
Lysine	3.7	4.3	4.8	4.5	7.0
Methionine	0.3	0.5	0.4	0.4	5.7
Cystine	1.0	1.2	2.0	1.3	–
Fenylalanine	3.5	3.9	3.5	3.6	9.3
Tyrosine	4.4	4.8	2.7	3.6	–
Threonine	3.9	4.1	3.5	3.6	4.7
Trytophan[2]	–	–	–	–	1.7
Valine	2.7	3.6	2.8	3.0	6.6
Nonessential					
Aspartic acid	10.6	9.5	10.2	9.7	
Glutamic acid	21.5	21.6	25.4	24.3	
Alanine	3.1	3.3	3.2	3.4	
Arginine	7.8	8.6	8.5	9.1	
Glycine	3.8	3.6	3.5	4.0	
Hystidine	–	0.8	3.2	–	
Proline	3.5	4.3	3.7	4.2	
Serine	5.3	4.9	5.0	5.5	

Adapted from reference 13 and 14.
[1] See reference 17.
[2] Not measured.

may have up to 20%. This is similar to soya and has 2–3 times the oil content of other legumes. Present genetic selection efforts are directed at obtaining seeds with higher oil yields. The species with the greatest genetic variability is *L. mutabilis*, thus research efforts are concentrated on this species. The fatty acid composition of several lupin species is summarized in table 3. Oleic acid is the predominant fatty acid for *L. albus* while linoleic predominates in *L. luteus* and *L. angustifolius*. The ω3, α-linolenic acid, content is much higher than in corn oil and is similar to that of soya oil. The saturated fatty acids are very low in lupin oil and there are measurable amounts of C_{20} monoenoic and C_{22} fatty acids. Erucic acid content is barely detectable, reported values in lupin oils are low enough to suggest no adverse toxic effects due to its presence. The low-saturated, high monounsatu-

Table 3. Fatty acid composition of sweet lupins
(% total methyl esters)

		Angusti-folius	Albus astra	Albus multolupa	Luteus aurea
Myristic	14:0	trace	trace	trace	trace
Palmitic	16:0	11.0	8.0	5.7	5.6
Palmitoleic	16:1	trace	trace	trace	trace
Stearic	18:0	5.5	2.4	1.3	1.6
Oleic	18:1	30.8	56.2	53.3	22.2
Linoleic	18:2	45.2	18.7	22.0	51.2
Linolenic	18:3	4.1	8.0	8.7	10.1
Arachidic	20:0	0.5	trace	–	1.6
Eicosaenoic	20:1	–	3.8	5.1	1.8
Behenic	22:0	2.9	2.8	3.8	5.8

Adapted from reference 18.

rated, and balanced ω6:ω3 fatty acids make the profile similar to that of soya oil and likely to be comparable in terms of health benefits [2, 5, 18].

Nutritional Value of Lupin Proteins for Humans

Traditionally, methods to assess protein quality have measured the effect of a single protein source on weight gain of malnourished infants or rapidly growing animals, such as the rat. This approach tends to underestimate the protein quality of plant proteins relative to animal foods. Since essential amino nitrogen needs are directly related to protein turnover rates, a given protein may support the growth of a normal child yet may be insufficient to meet the requirements for accelerated growth of the rat or of a child recovering from malnutrition. In addition, the possible complementarity of essential amino acids, observed when a legume-cereal mix is fed, is lost when single sources of proteins are studied. Quality of plant proteins measured by the traditional protein efficiency ratio (PER) can be underestimated for these two reasons. Most studies of lupin protein quality have thus far used the traditional assessment of net protein utilization and biological value of absorbed nitrogen [15–17].

Lupin protein quality has been evaluated in rats, pigs and humans. The PER in the rat has served to confirm that methionine is the limiting amino acid for its utilization. The PER will improve gradually when 0.1–0.4% *DL*-methionine is added to the lupin flour diet. The maximal PER value we have obtained with *L. albus* cultivar Multolupa protein is 1.3; after 0.4% *DL*-methionine is added, the PER raises to 2.3, which is slightly less than the 2.5 obtained with casein, used as a reference protein. The apparent protein digestibility of *L. albus* cultivar Multolupa measured in rats also improved from 75 to 80%, with methionine supplementation, relative to casein. In other studies using *L. angustifolius* cultivar Unicrop digestibility of 0.2% *DL*-methionine supplemented lupin protein was increased to 88% relative to casein [9–11, 13].

To test the safety of sweet lupins, Ballester et al. performed studies on rats consuming diets with 20% protein derived from *L. albus* and *L. luteus* supplemented with 0.3% *DL*-methionine. Growth of rats fed supplemented *L. luteus* was similar to that of the casein-fed control animals while those fed *L. albus* gained slightly less weight. No differences were found in organ-to-body weight ratio of liver, spleen, heart and adrenals after 3 months of feeding with the experimental diets. Gross autopsy and microscopic examination also disclosed no significant differences. Multigeneration studies in rats have confirmed that there were no adverse effects of feeding sweet lupins of these two species on growth, development, reproductive and lactational performance. No biochemical, histological or morphological alterations attributable to lupin were found after feeding it for three generations [10, 19].

The protein quality of *L. albus* cultivar Multolupa was evaluated by us in 8 young adult humans using the nitrogen balance technique at graded levels of N intake. Lupins were compared with whole dry egg protein. Lupin protein was fed at 0.4, 0.6 and 0.8 $g \cdot kg^{-1} \cdot d^{-1}$ and egg protein was given at the levels of 0.3, 0.45 and 0.6 $g \cdot kg^{-1} \cdot d^{-1}$. The levels of protein intake were randomly assigned using a modified Latin square. Energy intake was individually adjusted to meet individual needs. The apparent digestibility of the protein ranged from 70.2 to 78.8% for lupin and 67.1 to 83.2% for egg at the lower and higher levels, respectively. At equal protein intake level, digestibility of egg was significantly higher than that of lupin ($p < 0.05$) [15, 20]. López de Romaña et al. [21] have reported higher apparent protein digestibility values for *L. mutabilis* when fed to children at a level of 1.96 g lupin protein $\cdot kg^{-1} \cdot d^{-1}$. We have reported similar apparent digestibility in Chilean adult males for egg protein when evaluated at the same levels of intake [15]. A difference in apparent N digestibility, such as we observed between lupin and egg, is commonly found when animal protein is compared with vegetable protein. In the specific case of lupin, this difference in protein digestibility could be due to the presence of high levels of nondigestible polysaccharides such as stachyose and verbascose [14, 15, 20].

The results of the true nitrogen balance at different intake levels demonstrated that all subjects consuming the lupin diet at the level of 0.8 $g \cdot kg^{-1} \cdot d^{-1}$ protein were in positive N balance with a mean value of +16.4 mg. This value decreased to 0.2 mg N·kg^{-1}·d^{-1} for the level of 0.6 g protein and became markedly negative for the level of 0.4 g of protein·kg^{-1}·d^{-1} with a mean value of –15.1 mg N. Statistical comparison of the slopes obtained from individual regression equations of N intake versus N retention for both proteins showed a p value of 0.09, although this was not significant, the slope for egg was 30% higher than that of lupin [15, 20].

The relationship between N intake and N retention was linear and significantly positive in all subjects. The slope of the regression line of N intake on N retention is closely related to net protein utilization. Based on the comparison of the slopes, lupin has a protein utilization value of 77% relative to egg protein. Similarly, the regression coefficient of N absorbed versus N retention is a measurement of biological value. For this index, no significant differences were observed between both. Lupin protein had a biological value of 91% relative to egg protein [15, 20].

Sweet lupin flour (33%) blended with oat (25%) and wheat flour (16%) with minimal amounts of whey milk protein (5%) and sucrose (22%) has been used during the nutritional recovery of malnourished infants. Ten infants were given this mix and compared to a similar group that received a traditional cow's-milk-based formula with added carbohydrates. The recovery was similar in both groups in terms of weight gain, serum biochemical and nutritional indices, and feeding tolerance [22].

Additional research efforts have been directed at testing the incorporation of lupin flour to commonly consumed foods, such as bread, crackers or pasta [23, 24]. In a controlled trial in 31 adults, feeding tolerance, anthropometry and biochemical indices remained fully normal after 2 months of feeding a study cracker in which half the wheat flour was replaced by lupin flour. For 2 months, half the group received 150 g of lupin per day while the remaining were given an equivalent wheat cracker. At the end of this period the groups switched according to a double-blind design. The acceptability of both types of crackers was excellent at the outset. Tolerance was very good although some subjects had less gastrointestinal symptoms than others. These manifestations were attributable to satiety independent of the type of cracker being eaten. There were significant weight gains in both groups since the subjects continued consuming their usual diet [24]. Other studies incorporating lupin flour to traditional foods have confirmed these findings [25–27].

Significance of Sweet Lupins as Plantfoods for Humans

The main advantages of lupin relative to other legumes used in human nutrition relate to their high protein content; although deficient in sulfur amino acids, lupin protein is complementary to cereal proteins, thus the mix will be of higher

Table 4. Chemical composition of lupins and other legumes (g/100 g)

	Albus astra	*Luteus aurea*	*Albus multolupa*	Soya	Common bean	Lentils	Chick-pea
Kilocalories	333	256	352	403	317	326	349
Protein	37.9	42.5	36.7	34.1	20.6	24.0	18.2
Lipids	13.6	5.1	15.4	17.7	1.6	1.3	6.2
N-free extract	22.0	16.9	23.9	33.5	57.3	57.4	57.7
Digestibility,%	75	75	75	86	78	78	78
Amino acid score[1]	22	22	22	38	31	24	31

Adapted from reference 14 and 46.

[1] Percent of egg reference protein, methionine is first limiting for all legumes.

biological value. Lupins have twice as much protein as beans, chick peas, lentils and other legumes as shown in table 4 [2, 5, 28]. Sweet lupin seeds can be consumed directly after cooking or deep frying in a variety of culinary preparations. This has been the traditional way they are consumed in the Andean region. Today, in Italy and other Mediterranean countries they are served as snacks. Cooked lupins can also be made into spreads similar to 'hummus' or deep fried as in 'falafel' and consumed with salads or breads as done in the Middle East. Fermented lupin products similar to 'tempeh' have also been produced and tested for sensory properties with good results [2, 14, 29].

Lupins can be milled into flour and used in the preparation of multiple food products. The incorporation of 10% full-fat lupin flour, or higher levels if the flour is defatted, to wheat products such as pasta has a measurable effect on the amino acid score as illustrated in figure 2. The blend of 10–40% lupin flour to wheat flour will improve amino acid chemical score from under 40% to over 70% relative to egg reference protein. Since cystine can partially substitute for methionine, lysine becomes the limiting amino acid for the mix. In addition the lupin-wheat pasta made from the blend will acquire a yellow color that will persist after cooking, similar to egg-enriched pasta, enhancing consumer acceptability. Wheat flour based bread, crackers or cookies can also be made with up to 12% lupin flour, the nutritional benefit of these products is similar to that of lupin-enriched pasta [30–32]. Products of this type have been tested in several massive school-feeding programs in Chile and other Andean countries with excellent results in terms of acceptability [23, 25, 33].

Lupin flour can be defatted, this concentrated protein product (60–70% protein) blended with processed cereal flours and milk (lupin-wheat-milk) can be

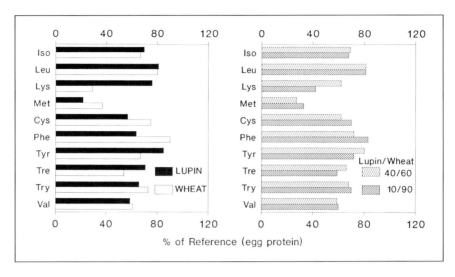

Fig. 2. Chemical amino acid score of *Lupinus albus* cv. Multolupa, wheat and lupin/ wheat blends relative to egg reference protein. This was calculated according to reference 17.

used for beverages or milk imitation products. These are less costly than milk and may use local agricultural products rather than imported dairy materials. Most developing countries have limited milk production capacities. Lupin protein isolates have also been prepared by alkaline extraction and acid precipitation. The testing of functional properties revealed better solubility than for soya isolates and similar emulsion capacity. Lupin may replace soybean protein isolates in many of their uses [2, 14]. Other additional benefits relate to the fatty acid pattern which is comparable to that of soy. Fiber content of the intact seed is high and will serve as a ready source of crude fiber for humans if consumed directly. Lupin hulls may also serve as a source of fiber for incorporation into food products [34, 35].

Potential risks are mainly related to toxic alkaloids present in the bitter lupins and to contaminant toxins produced by fungal infections of lupins. Lupinosis in sheep is caused by consumption of lupin greens which are infected with the fungal agent, *Phomopsis leptostromiformis*. The fungus produces potent hepatoxins called phomopsins. The development of phomopsin-resistant lupin strains promises to solve this problem [36, 37].

The alkaloids in lupin belong to a family of quinolein compounds called quinolizidine. The main ones present in lupin are sparteine, lupanine, hydroxylupanine, anagryne, cystisine, multiflorine, matrine and angustifoline. They are made solely by the green tissues of the lupin plant but are present in parts including the lupin seeds. They are present in higher amounts in seeds from plants that form a few large seeds rather than many small seeds. Understanding the reason for the presence of the quinolizidine alkaloids in lupins has proved elusive. It is highly unlikely that alkaloids are there by chance alone, it is most likely that they have contributed to the evolutionary fitness of the plant. Present thinking based on the concept of chemical ecology suggests that alkaloids are there to protect the plant from microorganisms, herbivores and competing plants. Sweet lupins are derived from homozygous mutants, since sweetness is a Mendelian recessive trait. Even after seeds are carefully selected there is a variable but high reversal rate to bitterness, thus suggesting that the genes responsible for alkaloid production are highly conserved. The experimental evidence available to date indicates that several alkaloids have specific plant antiviral, antifungal and antibacterial properties. The alkaloids also inhibit the germination of grass seeds and other weeds that grow in their vicinity. Alkaloids will deter snails and some insects from eating bitter lupin greens thus improving their selective survival at times of heavy pest infestation. In addition the alkaloids are toxic for vertebrates affecting the acetylcholine receptor and protein synthesis. They are mutagenic and teratogenic if ingested by pregnant animals. The concentration of alkaloids in green tissues increases within hours after experimental injury suggesting the quinolizidines are part of the plant's chemical defense system. It is noteworthy that some animals have learned to avoid consuming the bitter lupins, swine detect alkaloids at concentrations of 10^{-9}, other animals have evolved becoming resistant to the alkaloid effects. Given this evidence, the present strategy of breeding the alkaloids out of the plant may soon give way to alternate strategies such as cultivation of bitter varieties with postharvest processing of seeds or genetically modifying lupins to produce alkaloids in greens but preventing their translocation to the seeds [37–39].

From the standpoint of agricultural production, lupins represent one of the few legumes that can be grown successfully in temperate climates with cool nights. Lupins as opposed to other legumes can resist frost down to $-15\,^{\circ}$C and can grow with minimal use of fertilizers. They can be cultivated in extreme geographic conditions such as in Iceland, Alaska or the Patagonia in the Southern tip of the American continent. The nitrogen fixation properties of lupins is greater than that of other legumes, furthermore the citric exudates produced by the roots mobilize phosphorus and other minerals fixed in the soil. The inclusion of lupins in the rotation of crops by the ancient Andean cultures served not only to provide this protein-rich crop but also to significantly enhance soil fertility of cereals and tubers in the subsequent cultivations. They improve the quality of poor soils. The presence of alkaloids has been linked to improved resistance to insects and other pests, thus present efforts of genetic breeding include removing some of the alkaloids from the seeds while preserving them on the green parts of the plant. These naturally protected lupins may open the way to ecologically gentler and kinder agricultural practices since less fertilizers and pesticides are required. Based on these facts and their N fixation, P mobilization and climatic tolerance, lupins offer an excellent alternative for sustainable agricultural production with a positive impact on the environment [40–44].

Human survival and evolution have been largely determined by food availability. Protein supply of diets is a major determinant of an adequate nutritional status and overall well-being in humans. Plants provide the main source of food for human consumption, they supply not only dietary energy but also amino nitrogen for maintenance and growth. The human demand for protein is not equally satisfied by plant or animal food. Consumers globally are willing to pay higher prices for animal proteins, thus production of protein sources is dictated by the market forces rather than by ecological, biological or ethical considerations [45]. Consumer preference of animal protein is based not solely on nutritional quality but is subjective sensation related to cultural practices, food beliefs and habits established early on in life. Biological quality of proteins is poorly correlated to their market price, yet nutritional value is a necessary condition in gaining consumer preference over time. Consumer preference and not nutritional quality is what mandates what protein sources will be produced over time [15, 45]. Despite our best wishes and the many advantages of lupins we described, their incorporation in the human diet will be conditioned to the development of food products that will please the consumer. The challenge for human nutritionists and food technologists alike is to develop food products and preparations that will increase lupin consumption.

Acknowledgments

The careful review and suggestions of Dr. George Hill (University of Lincoln, New Zealand) is gratefully appreciated. The research collaboration of Dr. Digna Ballester, Dr. George Owen and Dr. Juan Ignacio Egaña is also acknowledged. The present work was carried out with the partial support of Fundacion Chile (Santiago, Chile), United Nations University World Hunger Programme (Tokyo, Japan) and Grant from the Chilean Fund for Scientific and Technological Research (FONDECYT).

References

1 Bressani R: Legumes in human diets and how they might be improved; in Protein Advisory Group of the United Nations Systems. Nutritional improvement of food legumes by breeding. New York, United Nations, 1973, pp 15–42.
2 Hill GD: The composition and nutritive values of lupin seed. Nutr Abstr Rev B 1977;47:511–519.
3 Aguilera JM, Trier A: The revival of the lupin. Food Technol 1978;32(8):70–76.
4 Gross R, Bunting ES: Agricultural and Nutritional Aspects of Lupins. Proceedings of the 1st International Lupin Workshop, Lima, April 1980. Eschborn, 1982.

5   Gladstones JS: Lupins as crop plants (abstract). Field Crop 1970;23(2):2–137.
6   Cherkalin NM: The main perspectives for lupine breeding in the All-Union Research Institute of Legume and Groat Crops, Orel (abstract 2). II Conferencia International del Lupino, Torremolinos, 1982.
7   Espinosa F, Chateauneuf R: El lupino en Chile: Conocimiento actual y perspectivas. Centro de Estudios del Desarrollo, 1987. Ser Mater, No 171. Santiago, CED.
8   Gross R, Von Baer E: Posibilidades de *L. mutabilis* y *L. albus* en los países andinos. Arch Latinoam Nutr 1977;27:451–472.
9   Yáñez E, Gattás V, Ballester D: Valor nutritivo de lupino y su potencial como alimento humano. Arch Latinoam Nutr 1979;29:510–520.
10   Ballester D, Yáñez E, Erazo S, López F, Haardt E, Cornejo S, López A, Pokniack J, Chichester CO: Chemical composition, nutritive value and toxicological evaluation of two species of sweet lupine (*Lupin albus* y *luteus*). Agric Food Chem 1980;28:402–405.
11   Yáñez E, Ivanovic D, Owen DF, Ballester D: Chemical and nutritional evaluation of sweet lupines. Ann Nutr Metab 1983;27:513–520.
12   Hove EL, King S, Hill GD: Composition, protein quality and toxins of seeds of the grain legumes glycine max, *Lupinus* spp., *Phaseolus* spp., *Pisum sativum* and *Vicia faba*. J Agric Res 1978;21: 457–462.
13   Ivanovic D: Perspectivas del lupino dulce en la alimentación humana en Chile. Estudio químico y nutricional del lupino dulce (*Lupinus albus* var. Multolupa); thesis, Santiago, 1980.
14   Yáñez E: Lupin as a source of protein in human nutrition. Proceedings 6th International Lupin Conference, Pucón, November, 1990, pp 115–123.
15   Uauy R, Yañez E: Plant foods for human protein nutrition: Studies on soy, lupin and mixed vegetable sources. Qual Plant Foods Hum Nutr 1983;33:17–28.
16   Pellet PL, Young VR: Nutritional evaluation of protein foods. UN World Hunger Programme. Food Nutr Bull Suppl 1980;4.
17   FAO/WHO/UNU Energy and protein requirements. World Health Organ Tech Rep Ser 1985; 724.
18   Masson L, Mella MA: Materias grasas de consumo habitual y potencial en Chile. Composición en ácidos grasos. Santiago, Editorial Universitaria, 1985.
19   Ballester D, Brunser O, Saitúa MT, Egaña JI, Yáñez E, Owen D: Safety evaluation of sweet lupine. II. Nine-month feeding and multigeneration study in rats. Food Chem Toxicol 1984;22:45–48.
20   Egaña JI, Uauy R, Cassorla X, Barrera G, Yáñez E: Sweet lupin protein quality in young adult males. J Nutr, in press.
21   López de Romaña G, Graham CG, Morales E, Massa E, McLean WC Jr: Protein quality and oil digestibility of *Lupinus mutabilis*: Metabolic studies in children. J Nutr 1983;113:773–778.
22   Soza G: Estudio de recuperación nutricional de lactantes con fórmula láctea enriquecida con productos farináceo mixto (base lupino dulce); in Fundación Chile (ed): Situación, Análisis y Perspectivas del Lupino en Chile. Reunión de Trabajo, Santiago, Dec 1977. Santiago, Fundación Chile, 1978, pp 95–96.
23   Mermoud S, Schneider D, Oyarguren F, Moller E, Quiñones A: Estudio de incorporación de harina de *Lupinus albus* en la alimentación normal de un grupo humano; in Fundación Chile (ed): Situación, Análisis y Perspectivas de Lupino en Chile. Reunión de Trabajo, Dec 1977. Santiago, Fundación Chile, 1978, pp 81–86.
24   Gattás V, Barrera G, Yáñez E, Uauy R: Evaluación de la tolerancia y aceptabilidad crónica de la harina de lupino (*Lupino albus* var. Multolupa) en la alimentación de adultos jóvenes. Arch Lat Nutr 1990;40:490–501.
25   Villegas CR, Vega M, Sifri H: Empleo de harina de lupino en la alimentación escolar en Chile (abstract). II Conferencia Internacional del Lupino, Torremolinos, 1982, p 21.
26   Carreño P, Urrutia X: Galletas enriquecidas con harina de lupino dulce (*Lupino albus* cv. Multolupa). Composición química, calidad biológica, evaluación sensorial y aceptabilidad; thesis, Santiago, 1982.
27   Villegas CR, Vega M, Sifri H, Mendoza N: Incorporacíon de harina de lupino en galletas fortificadas con hemoglobina (abstract No. 110). V Congreso Chileno de Nutrición y Alimentación, Concepción, Oct 1982.

28 Von Baer E: Comparative advantages of lupin. Proceedings 6th International Lupin Conference, Pucón, Nov 1990, p 141.
29 Gross R, Reyes M: Utilización del *Lupinus mutabilis* en la alimentación humana (abstract No. 90). II Conferencia Internacional del Lupino, Torremolinos, 1982.
30 Ballester D, Yáñez E: Valor nutritivo de harina de trigo enriquecida con harina de lupino (*Lupino albus* y cv. Multolupa). III Reunión del Lupino en Chile, Temuco, 1979, p 11.
31 Zacarías I, Ballester D, Yáñez E, García E: Pan con harina de lupino dulce (*albus* cv. Multolupa). Evaluación biológica y aceptabilidad (abstract). II Conferencia Internacional del Lupino, Torremolinos, 1982, p 22.
32 Ballester D, Zacarías I, García E, Yáñez E: Baking studies and nutritional value of bread supplemented with full-fat sweet lupin flour (*L. albus* cv. Multolupa). J Food Sci 1984;49:14–16.
33 Mermoud J: Funcionamiento hepático en individuos alimentados con lupino dulce (abstract No. 94). II Conferencia Internacional del Lupino, Torremolinos, 1982.
34 Ballester D, Saitúa MT, Villarroel P, Egaña JI: Nueva fuente potencial de fibra dietaria: Salvado de lupino dulce (*L. albus* cv. Multolupa). Rev Chil Nutr 1987;15:101–106.
35 Feldheim W: Use of lupins as sources of lipids and dietary fibre in human nutrition. Proceedings 6th International Lupin Conference, Pucón, Nov 1990, p 124–131.
36 Nally A: Identificación y patogenicidad de hongos aislados de semillas de lupino dulce (*Lupinos albus*) en almacenaje (abstract). II Conferencia Internacional de Lupino, Torremolinos, 1982, p 27.
37 Wink M: Plan breeding: Low or high alkaloid content? Proceedings 6th International Lupin Conference, Pucón, Nov 1990, pp 326–334.
38 Blanco O: Variabilidad genética del *Lupino mutabilis* dulce. Proceedings of the 1st International Lupine Workshop, Lima, April 1980. Eschborn, 1982, pp 33–47.
39 Gondran J: The diseases of the white lupin crops in France: Prevention possibilities. Proceedings 6th International Lupin Conference, Pucón, Nov 1990, pp 277–279.
40 Von Baer E: Adelantos en el cultivo del lupino en Chile. Proceedings of the 1st International Lupine Workshop, Lima, April 1980. Eschborn, 1982, pp 115–121.
41 Mora S: Epoca de siembra y distancia en el rendimiento y contenido de proteína y aceite en *Lupino albus*, cv. Astra, *Lupino luteus* cv. Aurea y cv. Weiko II. II Reunión del Lupino en Chile, Temuco, 1978, foll No. 12.
42 Sanhueza E, Vega E, Von Baer E: Ensayos exactos de época de siembra y fertilización de lupino en Traiguén (1977, 78). II Reunión del Lupino en Chile, Temuco, 1978.
43 Castro E: Lupino: Cultivo y utilización. Campesino 1978;109(8):24.
44 Calegari A: Residual effects of lupin (*Lupinus angustifolius* L.) on corn yield. Proceedings 6th International Lupin Conference, Pucón, Nov 1990, pp 283–287.
45 Ford R: The international market for lupins. Proceedings 6th International Lupin Conference, Pucón, Nov 1990, pp 142–152.
46 Schmidt-Hebbel H, Pennachiotti I, Masson L, Mella MA: Tabla de Composición Química de Alimentos Chilenos. Facultad de Ciencias Químicas y Farmacéuticas, Universidad de Chile Santiago, 1990.

Ricardo Uauy, MD, PhD, Casilla 138–11, INTA U of Chile, Santiago (Chile)

Simopoulos AP (ed): Plants in Human Nutrition.
World Rev Nutr Diet. Basel, Karger, 1995, vol 77, pp 89–108

....

Barley Foods and Their Influence on Cholesterol Metabolism

Graeme H. McIntosh, Rosemary K. Newman, C. Walter Newman

CSIRO Division of Human Nutrition, Adelaide, S.A., Australia;
Montana Agricultural Experiment Station, Montana State University,
Bozeman, Mont., USA

Contents

Introduction 89
Genetic Variation in Barley 91
Composition of Barley Grain 93
    Protein and Lysine 93
    Lipids 94
    Minerals and Vitamins 95
    Carbohydrates 95
Barley and Cholesterol Metabolism 98
Milling and Processing of Barley 103
Conclusion 105
Acknowledgments 105
References 106

Introduction

Barley is possibly the earliest cereal grain in recorded history and is mentioned several times in the Bible. Archaeologists have found nearly 9,000 year-old carbonized grains of barley in southern Egypt [1]. Evidence there indicates that the grain was ground or milled with stones. Other areas in Africa and Asia where barley was grown at an early date are Ethiopia, Iran, Iraq, China and Korea [2]. Barley was grown extensively in Scandinavia in the Bronze age, about 2000 BC [3]. Columbus is credited with introducing barley to the New World when he had supplies sent to Haiti, where one of his ships was being repaired [4]. Barley production of Colonial America was predominantly for beer brewing and later for livestock feeding [5]. It was the first cereal crop introduced into Australia, arriving with the first fleet and being planted in 1788 [6].

The acceptance of grains as food sources is related to culture and status. Within the last century, white bread became a status symbol, available only to the affluent when fine milling was first practiced. The choice of grain types for food seems to follow a socioeconomic pattern as well. Taylor [7] pointed out that historically, as nations have become prosperous, they progressed from barley to rye and then to wheat as the preferred grain for bread. There are still many countries, Tibet, Korea, Mongolia, Northern Africa and Asiatic countries where barley is the main staple for food use.

There is a need in the 1990s to recognize the importance of whole grains in our diet, for the contribution they make to our health and fitness. They supply significant sources of energy in the form of starches and protein, along with vitamins, minerals and complex non-starch polysaccharides (NSP), otherwise known as dietary fiber. The latter is important for a number of functions, including maintenance of normal gastrointestinal health, the prevention of colon cancer and heart disease.

There has been a tendency in the past to remove fiber because of its lack of food value, but this is now being recognized as inappropriate [8, 9]. Although not an essential nutrient, dietary fiber has profound effects on health. There is near unanimity amongst health authorities that the more economically developed countries should include more complex carbohydrate, starch and NSP, and less fat and sugar in the diet. Future research then seems likely to be focussed on dietary fibers likely to be protective against specific diseases.

Barley is the fourth most important cereal in the world in terms of world production (12% of total cereal production), and contributes to human nutrition in indirect (a main grain source for cattle, pigs and chickens), and more direct ways (beer, whisky, malted products and pearled barley). However, barley as currently represented on the supermarket shelf as pearled barley, is not fulfilling its potential in developed countries, being limited to use in soups and stews. In previous centuries it was a major food source in the form of breads, porridge or grits in countries such as Japan, Scandinavia and Scotland. One of the reasons barley became less popular was the greater preference for blander whiter food products (white = pure), and therefore wheat and rice took its place as the major cereal sources in several societies. Still today, white forms of pearl barley are much sought after, particularly in the Orient. However, the recognition that white rice and white bread are low fiber/low residue foods, and produce a rapid glucose rise in the blood, has been closely linked with knowledge of their tendency to promote, rather than protect from degenerative diseases. The high plasma cholesterol which correlates with heart disease incidence is recognized to be closely linked with inadequate dietary fiber intake on the one hand, and with specific types of dietary fiber on the other. In this respect, barley has been shown to be preferable to wheat and wheat products, by virtue of the character of its dietary fiber and its ability to lower plasma cholesterol via the fiber component β-glucan. β-Glucan is a close relative of starch, but is indigestible in the small intestine and is highly viscous when in solution. It is this characteristic which is considered to limit the availability of certain nutrients such as starch, fat and cholesterol, protein and minerals. β-glucan is also present in oats and oat bran, but barley is a better source, having potentially both a higher concentration, and a more uniform distribution throughout the grain.

Genetic Variation in Barley

Cultivated barley is a diploid, having seven pairs of chromosomes which control the expression of a wide range of physiological and morphological characteristics. Nutritional and malting quality of the kernel is determined by both genotype and environmental growing conditions. Thus, genetically controlled factors have a significant influence on nutritional quality and end-product utilization. As with malting quality, many genetic characters influence nutritional quality. It is relatively easy to select for a single quality character. However, it is difficult to achieve an acceptable level of numerous quality characters in a single plant.

Barley has three spikelets at each rachis node that are alternately placed. In six-rowed types, all three spikelets are fertile, whereas in two-rowed types only the central spikelet is fertile. Kernels from two-rowed barleys tend to be symmetrical, although kernel size may differ with position on the rachis. Larger kernels occur in the middle of the spike with smaller ones near the base and tip. The central kernels in six-rowed types are also symmetrical, but the lateral kernels are twisted due to spike attachment and crowding. Nutritional implications of two- versus six-rowed barley have not been investigated in great detail, but limited data suggest some nutritional advantages for two-rowed barleys [10]. It is quite possible that two-rowed barleys will have an advantage in milling and rolling because of the greater uniformity in kernel size and shape. This has not been investigated and will require an in-depth study to verify this hypothesis.

The most common or popular barley used for malting or animal feed is covered barley, in which the lemma and palea (floral parts) adhere to the caryopsis (grain) at maturity forming a persistent hull. For food consumption, the lemma and palea (or hull) of covered barley must be removed by abrasion or pearling, producing the familiar pearled barley. The aleurone, or layer outside the endosperm and beneath the testa and pericarp is also removed in 'heavy' pearling, resulting in significant losses of essential amino acids, oil and vitamins which are concentrated in this tissue.

Table 1. Effects of the hull-less and waxy gene on total, soluble and insoluble dietary fiber content in barley (% kernel dry matter)

Genotypes	Total dietary fiber		Soluble dietary fiber		Insoluble dietary fiber	
	N	W	N	W	N	W
CP/WP	18.97	19.97	5.18	6.42	13.81	13.55
NP/WNP	13.97	13.05	5.90	5.84	8.07	7.21
SNP/WSNP	12.51	15.59	4.45	7.85	8.06	7.74
B/WB	18.98	20.96	6.01	7.17	12.98	13.79
NB/WNB	13.08	13.51	5.94	6.90	7.14	6.63
SNB/WSNB	11.17	13.53	4.70	7.38	6.01	6.14
	14.78	16.10	5.36	6.93**	9.35	9.17

Data from Newman and Newman [12].

CP = Compana; WP = Wapana; NP = Nupana; WNP = Wanupana; SNP = Shonupana; WSNP = Washonupana; B = Betzes; WB = Wabet; NB = Nubet; WNB = Wanubet; SNB = Shonubet; WSNB = Washonubet; N = non-waxy barley; W = waxy barley.

** $p < 0.01$, comparing normal to waxy barley.

When the genetic trait of naked or nude *(nn)* caryopsis is present, the hull falls off during threshing, as it does in wheat and rye. The resulting dehulled barley is especially desirable as a food grain because it requires no pearling, thus retaining nutrients that would otherwise be lost in processing. Although grain yield per land unit may be about 10–12% less because of loss of the hull in the field, the ultimate grain volume is not subsequently decreased by the need for the pearling process.

Another important genetic trait of barley is that of waxy starch, produced by the recessive waxy endosperm character *ww,* wherein the starch is almost totally amylopectin. Additionally, waxy barleys, particularly when they are of the hull-less type, are consistently higher in β-glucans and total dietary fiber than non-waxy types [11].

Table 1 compares pairs of non-waxy and waxy starch barley genotypes for total, soluble and insoluble dietary fiber. The mean total dietary fiber for non-waxy types was 14.8% compared with 16.1% for the waxy types. Average soluble dietary fiber values were 5.4% and 6.9% for non-waxy and waxy respectively, which is significantly different ($p < 0.01$) [12].

Merritt [13] discovered a mutant barley derived from the variety Glacier in which the starch contained 44% amylose. This high-amylose trait has been transferred into a hull-less cultivar which is relatively high (7%) in β-glucans, similar to the waxy types [unpubl. data, Montana Agricultural Experiment Station].

Other important genetic traits that affect nutritional quality are length of awn, shrunken endosperm, β-glucan content, and high-lysine content. The major emphasis in barley breeding to date has been in total yield per land unit and malting quality. With the increasing interest in use of barley as a food grain, its wide genetic diversity provides the capability of meeting a variety of nutritional, functional and food ingredient needs, while maintaining acceptable agronomic characteristics.

Composition of Barley Grain

Studies on the chemical composition of barley have previously been conducted mainly by agronomists, animal scientists and brewing technologists, rather than by food scientists. Food composition tables generally include only pearled barley, the most common form utilized in the past. Although wide variations exist in certain components due to genotype and environment, barley seed structure and its basic components are relatively constant. Figure 1 is a diagram of a grain of barley, with enlarged areas showing the infrastructure of the seed.

The aleurone and subaleurone layers (fig. 1a), as well as the endosperm cell walls (fig. 1b) are major sites of dietary fiber and protein, while the endosperm (fig. 1b) contains the highest starch concentration.

Protein and Lysine

Barley typically ranges in protein content from 8 to 18%, with an average of 13%, which is about equal to all wheats and generally higher than most other cereal grains. The quality of the protein is relatively high, with protein efficiency ratio (PER) values averaging 2.04 for ordinary cultivars. High-lysine cultivars have been developed which contain 5.0–6.5 g lysine/100 g N compared with the normal lysine content of 3.0–3.5 g [3]. One of these high-lysine cultivars (GA700202) [14] produced PER values similar to milk protein [15] and has application in the livestock industry, to reduce or eliminate expensive protein supplements. More importantly, they have potential value to provide much needed quality protein for humans in developing countries. High lysine barley was used to produce Egyptian flat bread with improved protein quality [16]. Pedersen et al. [17] developed acceptable barley weaning foods from a new Danish high-lysine cultivar with improved energy and protein for infants in areas where animal protein foods are scarce or non-existent. More recently, a number of high protein mutants have been reported which have a 10% higher lysine content combined with good agronomic grain-yielding features [18].

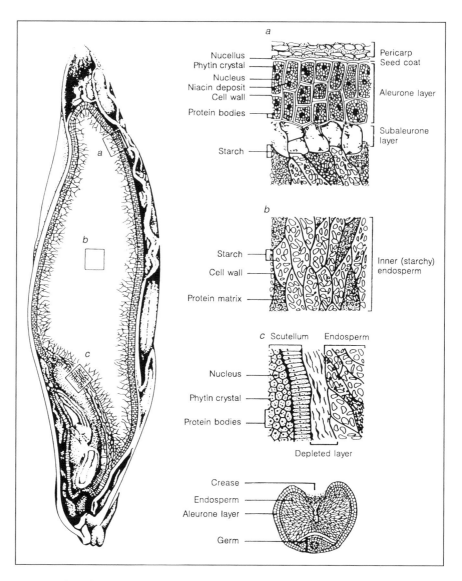

Fig. 1. Barley grain with enlarged cross-sections. Adapted from illustration by Wong and Fulcher [12].

Lipids

Barley generally contains from 2 to 3% oil, although genotypic variation exists, with some cultivars having up to 7% total lipid. The predominant fatty acids are linoleic (55%), palmitic (21%), oleic (18%), and α-linolenic (6%) acids. The lipid is distributed mainly within the aleurone and germ, and the relative concentration is subject to considerable genetic variability. Recent data [19] indicate a high percentage of oil in the pearlings of waxy, hull-less barley, which probably represents the aleurone layer and germ. Barley bran from this outer layer contained 6–12% oil.

The nonsaponifiable components of barley, representing about 8% of the oil, have received recent attention for their health-promoting properties. These include carotenoids, tocopherols, tocotrienols and isoprenoid products. Cholesterol-lowering activity has been identified in some of the E vitaminers (see later). A second cholesterol inhibitor in barley oil, α-linolenic acid, was identified by Qureshi et al. [20] in small concentrations, comprising about 6% of the total fatty acids.

Minerals and Vitamins

Mineral elements exist in similar concentrations in all cereal grains, and are essential to the health and growth of plants, as well as being important nutrient sources for animals and man. Crude ash content of barley ranges from 2 to 3%, with major components being phosphorus, potassium and calcium, with lesser amounts of chlorine, magnesium, sulfur, sodium and numerous trace elements [10]. Concentrations of minerals vary in the kernel, with the germ and aleurone layer having higher levels than the starchy endosperm. As with most cereal grains, phytate tends to irreversibly bind other minerals, particularly iron, zinc, magnesium and calcium. It may render these minerals biologically less available and create deficiencies when grain is a major component of the diet.

Barley germ contains all the known compounds with vitamin E activity. Qureshi et al. [20] identified d-α-tocotrienol as an inhibitor of cholesterol synthesis in the liver of experimental animals. The mechanism of its effect was reported to be suppression of HMG-CoA reductase, a rate-limiting enzyme in the acetyl-CoA-mevalonate pathway. Concentrations in the range of αT3:9.3 to 17.3 ppm were found in 9 American barley cultivars. In extracted barley oil, the following major vitaminers were: αT, 142 ppm, γT, 104 ppm, αT3, 558 ppm and γT3, 85 ppm. In concentration terms, αT3 accounted for 62% of the total vitamin E content [21, 22].

Barley is an excellent source of B-complex vitamins, especially thiamine, pyridoxine, riboflavin and pantothenic acid. High levels of niacin have also been reported in barley, but a portion of this vitamin is chemically bound to protein. It may be made biologically available by alkaline treatment. There are small amounts of biotin and folacin, and fat-soluble vitamins are limiting except for vitamin E.

Carbohydrates

Carbohydrates comprise about 80% of the barley kernel and provide the major source of available energy. Starch is the major carbohydrate and its concentration is inversely related to the total dietary fiber or NSP. A wide range in the percentage of total starch exists across genotypes, which is also influenced to a large degree by environment. Barley also contains small amounts of free sucrose, maltose and the trisaccharides raffinose, ketose and isoketose [23].

Starch in normal barley is predominantly amylopectin (74–78%), with the remainder being amylose. Waxy starch approaches 100% amylopectin, whereas the high-amylose starch contains only 56% amylopectin and 44% amylose. Uncooked normal, waxy, and high-amylose barley starches are highly susceptible to α-amylases. However, when high-amylose starch is autoclaved, a significant amount of resistant starch is produced. When fed to rats, relative glycaemia values for the autoclaved high amylose barley starch were significantly lower than those for wheat or waxy barley starch [CW Newman, unpubl. data]. Dietary fiber has been defined as the polysaccharides and lignin that are not digested by the endogenous secretions of the human digestive tract. Essentially this comprises the components from cell walls of plants and seeds. It also includes the polysaccharide food additives, now commonly found in the human diet, such as gums, algal polysaccharides, pectins, modified celluloses and modified or resistant starches. Cereals are good sources of dietary fiber, and in barley there is a relatively even distribution throughout the barley grain. Lignin is not a polysaccharide, but because of its indigestible nature it is included in dietary fiber. An unknown but variable component, resistant starch, acts like dietary fiber but is as yet ill defined in terms of its functional effects. Greater separation and definition will be required of fiber-rich foods and concentrates, in order to understand their contribution to human health.

The NSP of barley and other grains consist primarily of cellulose, β-glucan and arabinoxylans (pentosans) (table 2).

Cellulose is composed of linear $(1\rightarrow4)$-β-D-glucans, whereas the β-glucans, also linear $(1\rightarrow4)$ polymers, consist of β-D-cellotriosyl and β-D-cellotetraosyl residues separated by $(1\rightarrow3)$ linkages arranged in an independent or random manner [24]. Arabinoxylans of the starch endosperm are polysaccharides containing $(1\rightarrow4)$-β-D-xylan chains with α-L-arabinofuranose residues attached through $(1\rightarrow2)$ or $(1\rightarrow3)$ linkages. The β-glucans and arabinoxylans are concentrated in the aleurone and endosperm cell walls, and can be obtained in concentrated fractions by separation milling or air classification procedures.

The total dietary fiber of barley varies with genotype, as does the ratio of soluble to insoluble fiber. The waxy, hull-less cultivars are consistently higher in total and soluble dietary fiber, principally due to increased levels of β-glucans. When covered or hulled barleys are analyzed, the hull provides a high concentration of insoluble fiber. When the same barley is pearled, the proportion of insoluble fiber is lower, but the remaining portion of the kernel is still relatively high in fiber. The region from which the fiber comes within the grain also determines the ratio of soluble to insoluble fiber (fig. 2).

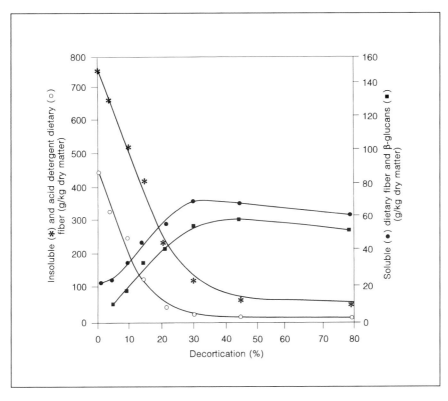

Fig. 2. Concentration of insoluble and soluble dietary fiber, cellulose and lignin (acid detergent fiber) and β-glucans in relation to the degree of decortication (g/kg dry matter). From Pedersen et al. [17].

Cereal type[a]	Dietary fiber			Total of NSP
	cellulose	β-glucan	pentosan	
Rye	1.5	2.5	8.9	13.9
Wheat	1.6	0.6	6.6	10.4
Barley[1]	0.4	4.2	6.6	12.0
Oats[2]	0.4	3.8	5.8	11.0

Table 2. Dietary fiber components of several cereal grains (g/100 g)

β-Glucan and pentosan have soluble and insoluble components, whereas cellulose (and lignin) are insoluble.

[1] Slightly pearled.

[2] Porridge oats.

There have been a number of claims for the biological benefits of brans as rich sources of dietary fiber, to inhibit a number of the degenerative diseases characteristic of western civilization [9]. Where some confusion exists is in the characterizing of dietary fiber in physicochemical and biologically functional terms, as the word 'bran' often refers to a poorly defined miller's fraction or by-product. Brans consist of a complex mixture of soluble and insoluble NSPs and other nutrient and non-nutrient components. Nor has the isolation of pure components such as cellulose always helped our understanding, as some modification of function may result from the processing. Nevertheless, numerous studies have now confirmed that effective cholesterol lowering can be achieved with foods enriched with dietary fibers from barley, particularly if the β-glucan-soluble dietary fiber is well represented.

Barley and Cholesterol Metabolism

Concern with control of cholesterol as a preventive health issue has become vitally important in modern society. The reduction of total dietary fats, and saturated fats in particular, is generally accepted as beneficial in serum cholesterol control. The introduction of acceptable functionally specific foods such as cereal grains with hypocholesterolemic properties can supplement the influence of dietary fat restriction. The total effects could be advantageous in terms of health costs, convenience and accessibility for the entire community.

It was initially shown in chickens fed barley, oats and pectin that significant cholesterol lowering could be achieved relative to wheat, corn, glucose and sucrose supplements [25]. In 1984, Burger et al. [26] showed that high-protein barley flour was very effective in lowering plasma cholesterol concentration in chickens. These workers attributed the effect to lipid-soluble factors, and subsequently identified them as d-α-tocotrienol and α-linolenic acid (see later). The studies of oats and cholesterol lowering have concentrated primarily on the soluble fiber components, particularly β-glucan [27, 28]. The β-glucan component of dietary fiber is well represented in barley, with many strains having higher concentrations than oats [29–31]. Consequently, the potential of barley and fractions of milled barley to achieve responses has been pursued in both animal model and human food studies. β-Glucan produces a highly viscous solution when solubilized in water, and initially this characteristic was tested against cholesterol-lowering ability. However, there was no good correlation, as pentosans containing xylose and arabinose are also capable of producing viscous solutions, and are well represented in the dietary fiber component of barley. Several studies [32–34] have demonstrated that enzymatic destruction of the barley β-glucan by pretreatment with β-glucanase reduces the viscosity, improves growth performance in chickens

Table 3. Selected barley studies showing evidence of plasma cholesterol lowering

Authors	Species	Cholesterol reduction, %	Comment
Burger et al. [26]	chickens	23	spent brewers grains
Qureshi et al. [20]	chickens	35	spent brewers grains
Fadel et al. [35]	chickens	13	waxy, hull-less barley
Klopfenstein and Hoseney [43]	rats	18.9	7% β-glucan for barley
Newman et al. [34]	chickens	25	waxy hull-less barley
McIntosh and Russell [39]	rats	14	barley vs. wheat
Newman et al. [40]	humans	3.3	barley vs. wheat
Newman et al. [41]	humans	5	barley and oats
McIntosh and Oakenfull [37]	rats	30	3% β-glucan from barley
McIntosh et al. [42]	humans	6	barley vs. wheat
Oakenfull et al. [44]	rats	52	barley vs. wheat
Martinez et al. [38]	chickens	47	various dietary fats

and partially or completely removes the cholesterol-lowering ability of barley. This helps confirm that the β-glucan is an important component of dietary fiber, responsible for growth inhibition and cholesterol lowering. The cholesterol lowering property of barley in experimental animal models has been well documented [35–37]. More recently, Martinez et al. [38] compared barley and wheat in chicken diets containing 1% cholesterol and 10% of palm oil, egg yolk, butter, tallow or corn oil. Chickens fed barley diets all had lower ($p < 0.0001$) total cholesterol levels compared to those fed wheat diets, regardless of fat sources. High β-glucan concentrates derived from milling or other methods have been tested for the cholesterol responses (table 3).

Mori et al. [45] roller-milled waxy hull-less barley and selected a milling fraction containing 13.6% soluble dietary fiber. This fraction was fed to rats, and serum cholesterol was significantly reduced ($p < 0.01$) compared with controls. Fecal dry matter was lower and fecal fat significantly higher in the barley fed rats. Kahlon et al. [46] fed to hamsters a diet containing 6% β-glucans from a milling fraction of barley which resulted in significantly lowered total plasma cholesterol compared with cellulose-fed animals. Similar reduced cholesterol levels were achieved with oat bran and rice bran diets. Newman et al. [47] treated waxy hull-less barley with β-glucanase and arabinoxylanase, in order to hydrolyze soluble fiber. The de-fibered barley did not lower serum cholesterol in hamsters, nor did cornstarch controls. However, animals fed the original whole-grain barley had significantly lower cholesterol. Klopfenstein and Hoseney [43] isolated and fed β-glucan from various grain sources including barley, and showed a significant cholesterol-lowering response in rats. Oakenfull et al. [44] isolated β-glucan from barley and fed it to rats. They titrated the β-glucan in diet to determine the concentrations required for maximum biological response, and it was found that 3% was as effective as 7% in achieving cholesterol reduction.

Barley has been shown to be equally effective in human lipid metabolism trials. Because wheat is a commonly used grain in bakery products and differs from barley in the proportion of soluble fiber, it has been used as a control product. Newman et al. [40], compared barley and wheat foods consumed by healthy volunteers for 4 weeks. Subjects who consumed wheat had slight but significant increases in total and LDL cholesterol when compared to pretreatment levels, while those who consumed barley had significantly lower cholesterol levels. This response was greatest in those with highest cholesterol levels initially. For a subsequent study with hypercholesterolemic individuals, Newman et al. [41] compared barley and oat foods. For both oats and barley groups, there was a lowering of serum total and LDL cholesterol, with no significant difference between the two treatment groups. McIntosh et al. [42] undertook a crossover study of 21 hypercholesterolemic men, in which barley and wheat foods of similar composition were fed for a 4-week period. Dietary fiber intake was increased by 17 g/day during the study and 8 g of β-glucan from barley was ingested daily, compared with 1.5 g during the wheat period. Total plasma cholesterol was lower by 6% and LDL by 7% as a result of barley foods. Plasma triglyceride and glucose were not significantly affected by the barley relative to wheat. A similar response (5.5% fall in total cholesterol) was produced when oat bran was used in a study involving otherwise the same comparisons and experimental approach [48] (table 4).

In a large oat and oat bran study, Davidson et al. [49] showed a significant response in cholesterol lowering (15.9%) with 56 g of oat bran containing 6 g β-glucan, confirming similar findings of others [50, 51, 28]. These studies show indisputable evidence that the β-glucan component of soluble dietary fiber in barley and oats is a useful agent for cholesterol control in hypercholesterolemic individuals.

Other components of grain have been examined for cholesterol-lowering effects. Welch et al. [52] separated and assessed independently in chickens the dietary fiber, protein and lipid fractions of oats. They were calculated to contribute 45, 30 and 16% to cholesterol lowering, respectively. From barley, the oil fraction appears to have variable cholesterol-lowering activity. Wang et al. [21, 22] demonstrated that barley oil lowered cholesterol in chickens, but not in hamsters [47]. The observation that hypocholesterolemic activity still exists in some barley cultivars that have been treated with β-glucanase suggests that other active components may be present [47].

There is considerable interest in the contribution that tocotrienols and α-linolenic acid may make to cholesterol lowering. There have long been claims that

Table 4. Effects of rice bran, oat bran and barley bran on plasma lipids in human studies

Relative to wheat bran	Wheat bran	Rice bran	Oat bran	Barley bran
Total cholesterol, %	100	98	94.4[1]	94[3]
LDL cholesterol, %	100	98.9	93.4[1]	93[3]
Triglyceride, %	100	91[2]	94	99

Data from McIntosh et al. [41] Kestin et al. [47].
[1] $p < 0.001$ oat vs. wheat.
[2] $p < 0.05$ Rice vs. wheat.
[3] $p < 0.05$ Barley vs. wheat.

vitamin E (of which *d*-α-tocotrienol is a member) has an influence on cholesterol metabolism [53]. Most studies have concentrated around α-tocopherol, the most biologically active form of the vitamin, but the tocotrienols are an interesting group of compounds with other important functional roles (e.g. antioxidants), and deserve further investigation. They are found in a number of oils (e.g.: rice oil, oaten and barley oil, palm oil), but there is some confusion currently as to whether α-, γ- or δ-tocotrienol is more biologically relevant to the cholesterol-lowering activity [20, 54]. It is proposed that tocotrienol has its influence on cholesterol metabolism by inhibiting HMG-CoA reductase enzyme, which is a rate-limiting factor in cholesterol synthesis. Studies in chickens [19] and humans [55] support the observation that there is a biologically active component in the extracted oil capable of lowering plasma cholesterol. Qureshi et al. [20] also identified α-linolenic acid as being biologically active in this respect. Although present in small concentration, this fatty acid is a member of the ω3 fatty acid synthesis pathway, best represented functionally by eicosapentaenoic acid in fish oils. These fatty acids are capable of important biological responses, including effects on lipid and cholesterol metabolism [56]. However, studies with linseed oil (a rich source of α-linolenic acid) on cholesterol metabolism in rats have produced little supportive evidence [M. Abbey, unpubl. data]. Oryzanol in rice bran oil is also under study for its influence on cholesterol metabolism. Because of the complex mixture of nonsaponifiable components in barley oil, further isolation and study of these factors is needed.

Barley spent grain, a by-product of the brewing industry (sometimes called barley bran flour), has application in food products. Brewer's grain is light brown in colour, is capable of being milled into a fine flour, and is suitable as a fiber

Table 5. Comparison of bran components from barley brewers grain (Miller Brewing Company, Milwaukee, Wisc., USA) and fractions

	Protein %	Total dietary fiber, %	Fat %
Brewers grain	25	50	8
Bran fiber	18	70	8
High-protein flour	35	35	8

supplement in baked products and breakfast cereals [57]. Table 5 shows protein fiber and fat content of brewers grain and two milled fractions.

The β-glucans are hydrolyzed and removed in the brewing process, leaving only insoluble fiber. However, tocotrienol and other lipid-soluble components remain, exerting an influence on lipid metabolism [58]. McIntosh et al. [59] compared brewers grain, wheat bran, two barley brans and cellulose in a rat feeding study. After 6 months, serum cholesterol in rats fed brewers grain was the lowest of all groups ($p < 0.05$), and 17% lower than the wheat bran group.

Robinson and Lupton [55] examined the effect of barley brewers grain flour and barley oil in 29 hypercholesterolemic men and women. Individuals consumed either 30 g of barley flour or 3 g of barley oil for 30 days, in combination with a low-fat diet. Serum cholesterol was significantly decreased with consumption of both the barley flour [–21.6 mg/dl] and barley oil (–18.7 mg/dl).

Proteins in barley have not been studied for their effects on lipid metabolism, despite evidence that high-protein barley flours are most effective in cholesterol lowering. Kritchevsky et al. [60] proposed that plasma cholesterol correlated with the lysine/arginine ratio of proteins. However, barley, oats and wheat have a very similar ratio (0.6). The hypothesis may be more applicable to animal/plant protein comparisons.

There has been considerable effort in defining the mechanisms whereby dietary fibers exert their influence in controlling lipid and glucose metabolism. There have been a number of reviews in this area [61, 62]. Progress has been hindered to a degree by the processes of isolation and purification of specific fiber fractions which can lead to loss of function. In general the most prominent theories for the mode of action of dietary fibers are:

(1) The viscosity effect, whereby lipids and other nutrients are inhibited from digestion and absorption by the presence of large NSP molecules in solution in the small intestine, (2) altered metabolism of bile acids (measured by the ratio of primary to secondary forms) and sequestering of bile acids, thus increasing their

elimination, (3) increased microbiological fermentation in the large intestine of polysaccharides leading to short-chain fatty acids some of which (e.g.: propionate) may feed back to the liver and inhibit cholesterol synthesis, (4) influence on insulin and glucagon secretion which are associated with lipid and carbohydrate metabolism. Both hormones are influenced by dietary fiber, with reduction in ambient concentrations in the circulation.

There is also some evidence that some grain components (e.g. wheat germ) contain factors which inhibit intestinal lipase activity and thereby alter lipid digestion and utilization [63].

Clearly there is evidence of the need for more research investigation in this area, to help characterize the nature of barley components and their influence on lipid metabolism. The importance of β-glucan and tocotrienols is that they provide at least two mechanisms whereby barley may effectively influence cholesterol metabolism.

Milling and Processing of Barley

Barley may be milled, pearled, flaked, malted, cracked, steel cut, crisped or roasted and polished to meet a variety of food- and beverage-processing needs. Blocking, a mild form of pearling, removes only the husk. Pearling, possibly the most common process currently used for the production of food barley, is an extension of this process, continuing until the pericarp, testa and aleurone layer have been removed. Well-filled, high-test-weight barley of uniform kernel size is most desirable for effective pearling. A technique has been developed in the oat milling industry for oat bran extraction and is being used in the production of barley bran, which consists of pericarp, germ, aleurone and subaleurone tissues. Barley provides a more concentrated source of dietary fiber for the food industry.

One type of pearling machine consists of 3–8 carborundum stones which revolve rapidly within a perforated cylinder. By this means the hull and bran layers are gradually rasped off, the degree of layer removal being determined by time. While the husk, pericarp and testa are relatively easily removed, the aleurone layer is more difficult. Considerable heat and pressure are involved in this processing, and Hayashi [64] and Nakumura [65] are of the opinion that the use of rice polishers provide a gentler means of removing these layers, resulting in higher quality bran and flours. Secondly, the serial layers are potentially concentrated sources of components such as phytate (P) vitamin E (including tocotrienols), vitamin C, carotenoids as well as β-glucans and proteins. The economics of this approach commercially speaking are needing critical appraisal.

Table 6. Representative chemical composition of waxy hull-less barley (average of Waxbar and Shonkin cultivars) and its milling fractions (% dry matter)

Milling fractions	Protein	Ether extract	β-Glucan	Dietary fiber		
				soluble	insoluble	total
Whole barley	14.2	2.6	5.8	5.5	7.1	12.6
Flour	11.9	2.0	3.1	1.7	2.7	4.4
Shorts	12.7	2.4	10.0	9.5	9.3	18.8
Bran	14.8	2.7	8.4	6.6	9.7	16.3

Data from Newman and Newman [12].

Although barley is not normally milled into flour and other fractions, as is commonly done with wheat, this can be successfully accomplished when appropriate milling conditions are met. Flour yields will vary with cultivar [66]. Bhatty [67] milled two hull-less barley cultivars and reported milling yields comparable to that of soft wheat. Ranhotra et al. [68] suggested that barley and its major milling fractions (bran and flour) may evoke different lipidemic responses and that by virtue of their higher soluble dietary fiber contents, they may be more effective in lowering blood lipids, than oat bran or flour. Mori [45] analyzed milling fractions of hull-less waxy barley and reported high concentrations of dietary fiber in shorts and bran fractions, with a marked hypocholesterolemic effect from the short fraction. Danielson et al. [69] reported increases in the concentration of total and soluble dietary fiber and in air-classified fractions of milled barley up to 36 and 20%, respectively.

Table 6 shows the mean chemical composition of two representative hull-less waxy barley cultivars before and after milling. The total dietary fiber is concentrated in the shorts and bran fractions, with higher β-glucan and soluble dietary fiber concentrations in the shorts.

Marlett [70] reported that the processing of a waxy hull-less and waxy covered barley into ready-to-eat cereal products increased the analytical solubility of dietary fiber. It was suggested that an observed reduction in the insoluble neutral sugar content in the fiber of these barleys was due to the pearling process. The total dietary fiber differed between unprocessed and processed barleys, (15.7 vs. 12.2% dry weight) although their total content of β-glucan was the same (5.1%). Heryford [71] and Fadel et al. [72] reported increased solubility of β-glucans when barley was extruded through a single screw-extruder. It is well known that the process of malting causes extensive changes in nutrient composition of barley

[10]. Barley malt and the concentrated water extract of malt are often used for flavoring or sweetening in baked products or ready-to-eat cereals. The effects of malting in general are a reduction in soluble and insoluble dietary fiber with an increase in free sugars.

Hull-less barleys, as used in the studies previously mentioned [45, 69, 70], are especially desirable as a food grain because they do not require pearling or blocking to remove the hull. This retains essential nutrients that would otherwise be lost due to these processing methods. A ready-to-eat breakfast cereal was prepared from a 50/50 mixture of waxy hull-less barley and whole wheat [73]. The grains were ground and blended with other ingredients to produce a 'crunchy nugget' product that was readily accepted by a trained taste panel. A 37-gram serving of this cereal provided 5 g of dietary fiber and only 1 g of fat.

This same barley was also milled through a roller mill to produce flour, bran and shorts. When the whole barley and the flour and shorts fractions were fed to rats, plasma cholesterol was significantly reduced compared to control rats fed a maize diet.

As previously stated, roller milling processes for barley have not been explored to any depths, especially as influenced by genotypes. Processes other than pearling have not been evaluated extensively for food product development, nor have their effects on nutrient content availability been investigated. This is an area that certainly needs more attention from food scientists.

Conclusion

Barley products have started appearing in the market place which need adequate definition and description, if representing fiber rich food sources. In this regard definitions are needed which will help protect industry and the public interest. In conclusion, newer approaches to milling and processing of barley provide potential for greater use of barley, in providing quality foods for human consumption. This should foster the increasing use of barley in western societies alongside other popular cereals.

Acknowledgments

We wish to acknowledge the financial support of the Grains Research and Development Corporation of Australia, Allgold of Australia (GHM) and the Montana Wheat and Barley Committee (RKN and CWN) with the research work undertaken in this area. It would not have been possible without the considerable technical and research student support we all enjoy. Preparation of the manuscript was aided by the CSIRO Sir Frederick McMaster Fellowship Scheme, which supported a visit to Australia by RKN and CWN.

References

1   Evans RD: Knossos Neolithic. II. Ann Br Sch Archeol (Athens) 1968;63:239–276.
2   Harlan JR: On the origin of barley; in Agriculture Handbook. Washington, Science & Education Administration, 1979, No 338: Barley – Origin, Botany, Culture, Winter Hardiness, Genetics, Utilisation, Pests.
3   Munck L: Barley for food, feed and industry; in Pomeranz Y, Munck I (eds): Cereals: A Renewable Resource. Theory and Practice. St Paul, AACC, 1981, pp 427–429.
4   Thacher JB: Christopher Columbus: His Life, His Work, His Remains. New York, Putnam, 1903.
5   Wiebe GA: Introduction of barley into the New World; in Agriculture Handbook. Washington, Science & Education Administration, 1979, No 338: Barley – Origin, Botany, Culture, Winter Hardness, Genetics, Utilisation, Pests.
6   Sparrow DHB, Doolette JB: Barley; in Lazenby A, Matheson GM (eds); Wheat and Other Temperate Cereals. Sydney, Angus & Robertson, 1975, pp 431–480.
7   Taylor A: Wheat needs of the world. J Home Econ 1918;10:11.
8   Temple JJ: Refined carbohydrates – A cause of suboptimal nutrient intake. Med Hypotheses 1983; 10:411–424.
9   Burkitt DP: Forward; in Kritchevsky D, Bonfield C, Anderson JW: Dietary Fiber Chemistry, Physiology & Health Effects. New York, Plenum Press, 1990, pp xi-xiii.
10 Newman CW, McGuire CF: Nutritional quality of barley; in Rasmusson DC (ed): Barley. Agronomy Monograph. Madison, American Society of Agronomy, Crop Science Society of America, 1985, No 26, pp 403–456.
11 Ullrich SE, Clancy JA, Eslick RF, Lance RCM: β-Glucan and viscosity of extracts from waxy barley. J Cereal Sci 1986;4:279–285.
12 Newman RK, Newman CW: Barley as a food grain. Cereal Food World 1991;36:801–805.
13 Merritt NR: A new strain of barley with starch of high-amylose content. J Inst Brew 1967;73:583.
14 Bang-Olsen K, Stilling B, Munck L: The feasibility of high-lysine barley breading – a summary. Barley Genetics V, 1991, pp 433–438.
15 Newman CW, Overland M, Newman RK, et al: Protein quality of a new high-lysine barley derived from Riso 1508. Am J Anim Sci 1990;70:279–285.
16 Newman RK, Nawar IA, Newman CW: High lysine barley improves nutritional value of flat breads. 13th International Congress of Nutrition, Brighton 1984.
17 Pedersen B, Bach Knudsen KE, Eggum BO: Nutritive value of cereal products with emphasis on the effect of milling; in Bourne GH (ed): Nutritive Value of Cereal Products Beans and Starches. World Rev Nutr Diet. Basel, Karger, 1989, vol 60, pp 1–76.
18 Aksenovich AV, Evtushenko EV: One possibility to increase the grain yield of lys 1/lys 2 barley. Barley Genetics VI, 1991, pp 439–440.
19 Wang L: Influence of oil and soluble fiber of barley grain on plasma cholesterol concentrations in chickens and hamsters; thesis, Montana State University, Bozeman, 1992.
20 Qureshi AA, Burger WC, Pederson DM, et al: The structure of an inhibitor of cholesterol biosynthesis isolated from barley. Biol Chem 1986;23:10544–10550.
21 Wang L, Newman RK, Newman CW, et al: Effect of barley oil on serum cholesterol in chickens. Cereal Foods World 1990;35:819.
22 Wang L, Newman RK, Newman CW, et al: Tocotrienol and fatty acid composition of barley oil and their effects on lipid metabolism. Plant Foods Hum Nutr 1993;43:9–17.
23 Henry RJ: The carbohydrates of barley grains. J Inst Brew 1988;94:71–78.
24 Bacic A, Stone BA: Isolation and ultrastructure of aleurone cell walls from wheat and barley. Aust J Plant Physiol 1981;8:453–474.
25 Fisher H, Grimminger P: Cholesterol lowering effects of certain grains and of oat fractions in the chickens. Proc Soc Exp Biol Med 1967;126:108–111.
26 Burger WC, Qureshi AA, Din ZZ, et al: Suppression of cholesterol biosynthesis by constituents of barley kernel. Atherosclerosis 1984;51:75–87.
27 Chen WJL, Anderson JW: Hypocholesterolemic effects of soluble fibers; in Vahouny GV, Kritchevsky D (eds): Dietary Fiber: Basic and Clinical Aspects. New York, Plenum Press, 1984, pp 287–306.

28   Shinnick FL, Mathews R, Ink S: Serum cholesterol reduction by oats and other fiber sources. Cereal Foods World 1991;36:815–821.

29   Henry RJ: A comparative study of the total β-glucan contents of some Australian barleys. Aust J Exp Agric 1985;25:424–427.

30   Henry RJ: Pentosan and $1 \rightarrow 3$, $1 \rightarrow 4$ β-glucan concentrations in endosperm and wholegrain of wheat barley oats and rye. J Cereal Sci 1987;6:253–258.

31   Åman P, Graham H: Mixed linked β-(1–3), (1–4)- D- glucans in the cell walls of barley and oats – Chemistry and nutrition, in Holmgren LK (ed): Symposium on Dietary Fiber with Clinical Aspects. Scand J Gastroenterol 1986;22(suppl 129):42–51.

32   Hesselman K, Åman P: The effect of β-glucanase on the utilisation of starch and nitrogen by broiler chickens fed on barley of low and high viscosity. Anim Feed Sci Technol 1986;15:83–93.

33   Graham H, Lowgren D, Pettersson D, et al: Effect of enzyme supplementations on digestion of a barley/pollard-based pig diet. Nutr Rep Int 1988;38:1073–1079.

34   Newman RK, Newman CW: Beta-glucanase effect on the performance of broiler chickens fed covered and hull-less barley isotypes having normal and waxy starch. Nutr Rep Int 1987;36:693–699.

35   Fadel JG, Newman RK, Newman CW, et al: Hypocholesterolemic effects of beta-glucans in different barley diets fed to broiler chickens. Nutr Rep Int 1987;35:1049–1058.

36   Newman RK, Newman CW, Fadel J, Graham H: Nutritional implications of β-glucans in barley. Barley Genetics V, 1986, pp 773–780.

37   McIntosh GH, Oakenfull D: Possible health benefits from barley grain. Chem Aust 1990;57:294–296.

38   Martinez VM, Newman RK, Newman CW: Barley diets with different fat sources have hypocholesterolemic effects in chickens. J Nutr 1992;122:1070–1076.

39   McIntosh GH, Russel GR: The role of barley in human nutrition; in Sparrow DHB, Lance RMC, Henry RJ (eds): Alternative End Uses of Barley. Parkville, Royal Australian Chemical Institute, 1988, pp 49–54.

40   Newman RK, Lewis SE, Newman CW, et al: Hypocholesterolemic effect of barley foods on healthy men. Nutr Rep Int 1989;39:749.

41   Newman RK, Newman CW, Graham H: The hypocholesterolemic function of barley β-glucans. Cereal Foods World 1989;34:883–886.

42   McIntosh GH, Whyte J, McArthur R, et al: Barley and wheat foods influence on plasma cholesterol concentrations in hypercholesterolemic men. Am J Clin Nutr 1991;53:1205–1209.

43   Klopfenstein CV, Hoseney RTC: Cholesterol lowering effect of β-glucan-enriched bread. Nutr Rep Int 1987;36:1091–1098.

44   Oakenfull DG, Hood RL, Sidhu GS, et al: Effects of barley and isolated barley β-glucans on plasma cholesterol in the rat; in Martin DJ, Wriley CWV (eds): Proceedings Cereals International Conference, Brisbane, 1991, pp 344–349.

45   Mori T: Chemical characterisation and metabolic function of soluble dietary fiber from select milling fractions of a hull-less barley and its waxy starch mutant; MS thesis, Montana State University, Bozeman, 1990.

46   Kahlon TS, Chow FI, Knuckles BE, Chiu MM: Cholesterol-lowering in hamsters by barley fractions, rice bran and oat bran, and their combinations (abstract No. 835). FASEB J 1992;A1080.

47   Newman RK, Wang L, Danielson AD, et al: Effects of barley soluble fiber, barley oil and rice bran oil on cholesterol metabolism in hamsters. FASEB J 1992;6:A1080.

48   Kestin M, Moss R, Clifton PM, et al: Comparative effects of three cereal brans on plasma lipids, blood pressure and glucose metabolism in mildly hypercholesterolemic men. Am J Clin Nutr 1990; 52:661–666.

49   Davidson M, Dugan LD, Burns JH, et al: The hypocholesterolemic effects of β-glucan in oatmeal and oatbran. JAMA 1991;265:1833–1839.

50   Anderson JW, Story L, Sieling B, et al: Hypocholesterolemic effects of oat bran or bean intake for hypercholesterolemic men. Am J Clin Nutr 1984;40:1146–1155.

51   Anderson JW, Spencer DB, Hamilton CC, et al: Oat-bran cereal lowers serum total and LDL cholesterol in hypercholesterolemic men. Am J Clin Nutr 1990;52:495–499.

52   Welch RW, Pederson DM, Schranka B: Hypocholesterolemic and gastrointestinal effects of oat bran fractions in chickens. Nutr Rep Int 1988;38:551–562.

53   Kibata M, Schimu Y, Miyake K, et al: Studies on vitamin E in lipid metabolism; in de Duve D, Hayashi O (eds): Tocopherol, Oxygen and Biomembranes. 1978, pp 283–295.

54   Qureshi AA, Chaudhary V, Weber FE et al: Effects of brewers grain and other cereals on lipid metabolism in chickens. Nutr Res 1991;11:159–168.

55   Robinson MC, Lupton JR: The effects of barley flour and barley oil on hypocholesterolemic men and women. J Am Diet Assoc, in press.

56   Simopoulos AP: Omega-3 fatty acids in health and disease and in growth and development. Am J Clin Nutr 1991;54:438–463.

57   Chaudhary VK, Weber FE: Barley bran flour evaluated as dietary fiber ingredient in wheat bread. Cereal Foods World 1990;35:560–562.

58   Qureshi AA, Schnoes HK, Din ZZ, et al: Determination of the structure of cholesterol inhibitor II isolated from high-protein barley flour. Fed Proc 1984;43:1866.

59   McIntosh GH, Jorgensen L, Royle P: The potential of insoluble dietary fiber from barley to protect from DMH- induced intestinal tumors in rats. Nutr Cancer 1993;19:213–221.

60   Kritchevsky D, Tepper SA, Klurfeld M: Dietary protein and atherosclerosis. JAOCS 1984;64: 1167–1171.

61   Anderson JW, Deakins DA, Bridges SR: Soluble fiber hypocholesterolemic effects and proposed mechanisms; in Kritchevsky D, Bonfield C, Anderson JW: Dietary Fiber Chemistry Physiology and Health Effects. New York, Plenum Press, vol 25, pp 339–363.

62   Schneeman BO: Gastrointestinal response to dietary fiber. Adv Exp Med Biol 1990;270:37–41.

63   Lairon D, Borel P, Termine E, et al: Evidence for a proteinic inhibitor of pancreatic lipase in cereals, wheat bran and wheat germ. Nutr Rep Int 1985;32:1107–1113.

64   Hayashi S: Alternative barley end use through methodological revolution; in Sparrow DHB, Lance RMC, Henry RJ (eds): Alternative End Uses of Barley. Parkville, Royal Australian Chemical Institute, 1988, pp 105–108.

65   Nakamura H: Human food uses of barley in Japan; in Sparrow DHB, Lance RMC, Henry RJ (eds): Alternative End Uses of Barley. Parkville, Royal Australian Chemical Institute, 1988, pp 93–98.

66   McGuire CF: Barley flour quality as estimated by soft wheat testing procedure. Cereal Res Commun 1984;12:53–54.

67   Bhatty RS: Physiochemical and functional (breadmaking) properties of hull-less barley fractions. Cereal Chem 1986;63:31.

68   Ranhotra GS, Gelroth JA, Astroth K, et al: Relative lipidemic responses in rats fed barley and oat meals and their fractions. Cereal Chem 1991;68:548–551.

69   Danielsen AD, McGuire CF, Newman RK, et al: Production of high fiber barley fractions. Barley Genetics VI. Helsingborg, 1991, pp 455–457.

70   Marlett JA: Dietary fiber content and effect of processing on two barley varieties. Cereal Food World 1991;36:576–578.

71   Heryford A: The influence of extrusion processing on the nutritional value of barley for weanling pigs and broiler chickens; MS thesis, Montana State University, Bozeman, 1987.

72   Fadel JG, Newman CW, Newman RK, Graham H: Effects of extrusion cooking of barley on ileal and fecal digestibilities of dietary components in pigs. Can J Anim Sci 1988;68:891.

73   Newman CW, Newman RK: High fiber barley for breakfast cereals; in Holas J (ed): ICC Symposium. Cereal-Based Foods: New Developments, Prague 1991, pp 216–223.

Graeme H. McIntosh, CSIRO Division of Human Nutrition, Kintore Avenue,
Adelaide, S.A. (Australia)

Simopoulos AP (ed): Plants in Human Nutrition.
World Rev Nutr Diet. Basel, Karger, 1995, vol 77, pp 109–134

....

The Nopal: A Plant of Manifold Qualities

Miriam Muñoz de Chávez, Adolfo Chávez, Victoria Valles,
José Antonio Roldán

Instituto Nacional de Nutrición S.Z., Community Nutrition Division, Tlalpan,
México, D.F., México

Contents

Introduction 109
    General Background 110
    Environmental Aspects 115
    Cultural Issues 116
Morphological Characteristics 118
Nutrient Content 119
    Proteins and Fats 119
    Vitamins 122
    Minerals 123
    Carbohydrates 123
    Organoleptic Qualities 124
Other Health and Nutritional Characteristics 125
Production and Consumption 127
    Production 127
    Consumption 130
Conclusions 132
References 132

Introduction

The plant we will discuss is an unusual one, not only because of its appearance, which is unique, but also because of its manifold applications in both animal and human nutrition. People eat its fleshy stems (cladodes), which look like leaves. The tender young cladodes that are at the ends of the plant, are harvested as vegetables. The fruit of the nopal, the prickly pear, is eaten raw after being peeled. The fermented juice of the fruit, without the seeds, is drunk, as a beer called 'colonche'. Moreover, a preparation is made by grinding the fruit, with the seeds, and boiling it until it is semi-dry and has the consistency of cheese: this is the prickly pear cheese.

From the standpoint of nutrition, the nopal has a twofold importance. On the one hand, it has valuable nutrients, from the vegetable (cladodes) and the fruit (prickly pear). On the other hand, and this is what makes it a unique plant, there are increasing data showing some metabolic effect of the cladodes that suggest that it can be even more beneficial than that of the 'best' popular vegetables. Probably most of these beneficial effects are due to the fiber, mainly to the soluble fractions that give to the vegetable a mucilaginous or slimy appearance. It is widely accepted, with some supporting data, that the cladodes cooked as vegetables have some beneficial effects. For example, they help reduce serum cholesterol level, regulate blood sugar, and control excessive gastric acidity. These effects are important to consider in this modern world full of risks stemming from food, overweight and stress.

As we describe this wonderful plant, almost sacred for Mexicans, we will discuss many elements among which we have to discern facts from fantasies. Those that believe in this plant, that live close to it and eat it often, assure that it is a real source of health. We will present some data on what the plant actually does for the nutrition and health of the poor and summarize some recent data that show the benefits to other population groups as a new vegetable to enrich the diet, perhaps with the consideration that it will not be just another vegetable to add to the long list of those already offered, but one of outstanding quality.

General Background

The nopal (*Opuntia* spp.) is a succulent originating in the American Continent. Taxonomically it belongs to the Cactus family, *Opuntia* genus, and Plantyopuntia subgenus. It grows abundantly in North American arid and semiarid areas. Actually, it is a very strange plant, different from any other in the world. Its shape is quite uncommon, the so-called leaves (cladodes) are actually stems that grow one on top of the other in a very irregular manner, giving the plant its very particular appearance (fig. 1). Due to its thorns, some big and visible and some small (but which can cause local irritation), it can be aggressive when one does not know how to handle it. And yet, the nopal has been an unlimited source of flavors, textures, and nutrients for those living close to it within its own ecological environment.

The nopal is almost essential to the Mexican countryside. It appears on the Mexican flag, on the Mexican Coat of Arms (fig. 2) and on most Mexican coins.

Fig. 1. The nopal with budding cladodes.

Fig. 2. The Mexican national flag.

The legend of how the Mexican nation was founded goes back to a group of Nahuatl pilgrims, to whom the gods gave a sign to find a place where they should finally settle. This place would be indicated by the presence of an eagle standing on a nopal devouring a snake (fig. 3) [1]. Surely that was a symbol pointing out the availability of appropriate conditions for the survival of the group. This legendary marked nopal was found on an island in the middle of a fertile lake where Mexico City, the largest city in the world, is located today. It was originally named Tenochtitlan, meaning the place of the nopal [2].

Fig. 3. Stone carving of the legendary eagle perched on a nopal.

Local artists have left numerous representations of *Opuntia* spp. such as the Tenochtli, a stone carving of the prickly pear. The *Codex Mendoza* (1549) depicts another beautiful hieroglyphic: the Tenochtli or divine prickly pear (fig. 4) [3].

When Franciscan monks established the first missions in California, they planted some nopal. There were two related species – *Opuntia ficus-indica* and *Opuntia megacantha.* The name of 'Indian fig', first known in California and later in Arizona, New Mexico, and Texas, was originally given to the former species [4]. Later, the name was applied either to all nopal or only to their fruit.

'... Missionaries and other settlers found these cactus plants were useful not only for their fruits but also as a major source of a mucilaginous material used to glue adobe bricks to build the missions ... Over time, both species of these cactus plants were also planted in large ranches, in hacienda main houses and surrounding the houses of peons and other dwellers. Thus, wherever Spaniards or Mexicans lived, plantations appeared, the practice being followed later by North American settlers [5].'

From Mexico the nopal was also shipped to Europe. The route followed is unknown, but it was planted in Portugal and southern Spain. From Spain it got to Tunisia, Algeria and Morocco, where, although the fleshly leaves are not eaten, the fruit is greatly enjoyed. In these Arab countries, this fruit is known under the

Fig. 4. Hieroglyphic from the *Codex Mendoza* (1549) depicting the legend of the nopal.

French name 'figue de barbarie'. The Arabs, however, called it 'camus ensara' meaning 'European fig'. In France, in addition, it is also known as 'opence', probably from the Latin 'opuntia'. The origin of the name 'opuntia' is unknown, but it could be derived from the Greek city Opontio. That city could have been the place where the plant flourished and perhaps it reached some of the Arab countries [6]. Whatever the route, the truth is that this plant is grown in a great number of countries around the Mediterranean where it is mainly used to reduce soil erosion in arid parcels, although, as mentioned, there are many regions in which the fruits are eaten.

The time when the nopal cladodes began being cultivated for human consumption is unknown. There is some discussion whether the old communities were placed in the middle of good wild nopal patches (nopaleras) or whether the communities selected and cultivated them. The most probable is the latter, because of many of the 'nopaleras', or patches of nopal forests seem not to be really wild, but screened either for the quality of their fruit or for the softer fleshy leaves with less aggressive thorns. These nopal plantations must have existed prior to the arrival of the Spaniards to Mexico.

The existence of so many different varieties, with fruits of different colors and flavors, and various types of fleshy leaves, varying from very fibrous and wild

Fig. 5. Nopal at fruit bearing season of the year. Prickly pears are abundant in season.

ones to the most tender and least thorny, could be evidence that humans have practiced plant selection very early.

The fruit is known as the prickly pear (fig. 5). Their tastes, textures and colors are quite variable. They can be various shades of red, yellow, green, purple or orange. The rural people can tell from which community the fruit comes from its appearance and taste. In the Central and Mid-Northern regions of the country, each town 'owns' a certain nopal variety producing the prickly pears of their choice. Each town assures theirs is the best tasting.

At present, only three countries produce nopal for commercial purposes: Mexico, Chile and Italy (mostly Sicily) for the prickly pears, and only Mexico produces cladodes as a vegetable [7].

The development of large commercial plantations in the central highlands of Mexico increased during the 1940s and stimulated agricultural research. In recent years, many changes have occurred in the quality and technology of nopal culture. Within a very short period of time, many results were obtained. Perhaps the great variety of existing types made it easy to improve the yield and the quality of the species to produce the vegetable. For instance, a thornless variety was developed as a fodder, although it was soon noticed that country rats ate them before they even grew [8]. This variety is now used as fodder for the maguey rat *(Neoloma ambigula)* which is very appreciated as food in some regions for its white and tasty meat.

The most recent development was the introduction of plants with very low and tender cladodes into some central areas in the country. They are now intensively produced to supply Mexico City markets and for export to the US.

Environmental Aspects

The nopal is a cactus that at present can be seen growing in arid and semiarid regions of the world although it may develop in various climates and soils. However, its optimum growth occurs in arid sandy land, derived from lime and igneous rock with a neutral or slightly alkaline pH. It can survive under extreme temperatures ranging from less than 0 to 50°C, the most common case in the plantations in the north [5].

Fifty-two percent of the Mexican land surface, that is about 10^6 km^2, has good characteristics for the different species of nopal, a fact that contributes to its wide geographical distribution. Some large expanses in the arid regions, unsuitable for other kinds of culture, are potentially available for plantations of some of the new commercial varieties.

Wild nopal species and varieties are well adapted to the climate and soil conditions where they grow. Some species, such as *Opuntia stricta,* may develop at sea level, while others, such as *Opuntia streptacantha* (giant cactus), grow easily at altitudes of 2,700 m or more above sea level (fig. 6).

The total number of species is estimated by some authors at 258, although only 110 different species are well defined. Of these, 65 are found in the Valley of Mexico [9]. All species grow wild, and only 8 of them, with their varieties, are used for human consumption. Among them, *Opuntia ficus-indica* is the most cultivated for fruit production, with 5 varieties preferred by the city market (table 1).

Although most cactus plants are immune to long droughts, the commercial varieties require annual rains to replace water gradually lost by their tissues. When no rain falls, stems become thinner, wrinkle, and cladodes and branches start to fall. If droughts continue for more than 2 years, the plants may die. An additional problem is competition for water and nutrients with other cactus species that may have a greater uptake capacity.

Wild species such as *Opuntia tormentosa* and *Opuntia robusta* depend on special pollination systems for their reproduction. Pollination is performed by certain bees such as those of the *Megachilidae* genus, although sometimes common *Apis mellifera* bees are also involved. Cross-pollination, from the point of view of evolution is of great advantage because it allows for a high degree of adaptability to environmental conditions [9].

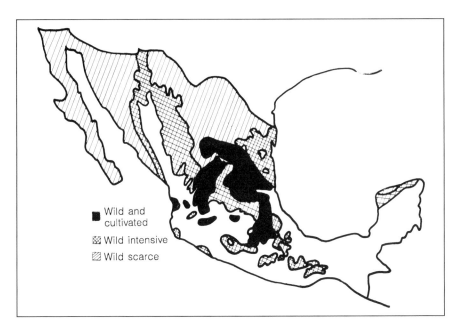

Fig. 6. Geographic distribution of wild and cultivated nopal in Mexico.

Seed spreading is made by animals, especially by fruit-eating birds. Since the seeds of the prickly pear have a hard cover, they pass through the digestive tract and travel variable distances before being excreted, sometimes in places very far away from the place where the seeds were eaten. This has greatly contributed to the geographic distribution of the nopal in Mexico.

Cultural Issues

In spite of current commercial interests, most nopaleras are still mostly linked to the life of the farmers of the arid high plateau of Mexico. Not only do they use the nopal as food, at times as a vegetable and at times a high-caloric and thirst-quenching fruit, but for many other purposes as well: for example to control erosion, fix boundaries of plots and protect their domestic animals; as fodder, as a construction material, and as adobe glue [10]. It is used to support the production of the nopal rat, a source of animal protein. In this rural culture, the nopal is also used as medicine and as ceremonial food. Nopal is also the best food for certain beautiful hummingbirds with long beaks and special flying which allows them to drink the flower nectar without getting hurt by the thorns.

The nopal is a distinctive host of the 'cochineal insect', a dipteron which reproduces in the fleshy leaves. Already in pre-Hispanic times, it was found that

Table 1. Characteristics of the main commercial varieties of prickly pears

Commercial names	Color		Crop season	Weight total g	% weight seeds	Seeds n	Edible portion %	Brix degrees
	cover	pulp						
Reyna	green-yellow	white	Jul–Aug	140	3.5	288	62	15.0
Cristaline	green-yellow	white	Aug–Sept	216	2.6	357	54	15.4
Chapeada	rose-yellow	white	Jul–Aug	133	3.8	296	51	14.2
Tabayeco	yellow-brown	white	Sept–Oct	145	3.0	249	48	15.9
Centenario	orange	orange	Aug	112	3.5	243	64	14.7
Bold red	green-red	red	Aug	184	3.0	279	61	15.1
Cardona	purple	purple	Jul–Sept	83	3.6	177	46	15.5

by milling the adult insect and cooking it, it changed its color to the most brilliant and beautiful red which no aniline has been able to reproduce. This scarlet dye was used by old Mexicans to color textiles, dye sculptures and buildings, and paint murals and codices. Today it is used to dye some of the most sumptuous military jackets in the world, such as those of the queen of England's personal guards, and many Persian and other rugs famous for their coloring. In the second most spoken language in Mexico after Spanish, Nahuatl, this dye is known as 'Nocheztli', meaning 'nopal blood' [11].

In the early 17th century, King Philip III stated that one of the most precious goods produced in the West Indies was the 'scarlet' dye derived from the cochineal insect valued at the time as much as gold. Between 1784 and 1789, 15,852 pounds of scarlet dye were produced in only one town, Xamiltepec. Before, natives of many communities like Otumba, Cholula, Tepeaca, Huexotzingo, and Tlaxcala destroyed their own nopaleras so as not to become slaves of 'the powerful mayors', that produced and exported the scarlet dye.

This fabulous product was gradually replaced by anilines, and the cochineal industry became almost extinct. Fortunately, in the last few years, anilines in turn are being replaced by natural products because of the many reports about allergens in cosmetics, drugs and foods [12]. The use of the cochineal and its commercial production are being reevaluated and some production is starting, although not only in Mexico, but also in other countries such as Peru and Spain (Canary Islands).

Traditionally, the tender, young cladodes that grew during the previous year are the ones preferred to eat. The flowers (of a very short life span) are delicious in soup or cooked with chili sauce, and the prickly pears are loved for their sweet, scrumptious taste and for their thirst-quenching quality in dry, hot areas. Older cladodes are used as fodder, and the most central and fibrous cladodes are used as fuel to cook the traditional tortillas and beans.

One could say that one almost must be Mexican to achieve so many varied uses of the nopal. One needs to have learned since early childhood how to sort through the many defenses of the plant to reach its many 'delights'. Fortunately, the new varieties of today, including dwarf plants that grow fast and give fleshy leaves, are easier to handle, and are the ones that are reaching the tables of some developed countries. Although concurrently, the new nopal culture is promoting an increasing loss of cultural habits linked with the integral use of the nopal.

Morphological Characteristics

The nopal is a succulent cactus plant consisting of a stem divided into fleshy parts. When buds emerge at the end of a cladode they are cylindrically shaped as small horns for a few days until the fleshy cladodes acquire their shovel- or racket-like shape, which in turn produce new small horns that soon become atrophied, leaving thorns in their place. The new cladodes perform most photosynthesis and store most water.

Flowers burgeon from areolae located on top of the fleshy leaves. Each areola produces flowers that bloom at different times. Some appear one year, others the following year, and still others, the third year. Petals are very bright and of many colors such as yellow, orange, pink, red, salmon, etc. depending not only on the species but also on the varieties. Natural flowering occurs in spring (March through May) although there are some nopaleras in certain parts of the country where flowering occurs during other seasons. Flowers are anatomically hermaphrodites although some are unisexual due to androecium atrophy. When the perianth is fertilized, it withers and falls.

The root holding the stem has also certain pecularities although it is similar to those of other dicotyledons. Because of their shape, they are typical roots with primary axes used to firmly set the plant into the soil; they are thick, not succulent, and their shape is proportional to the size of the stem. A root feature allows the nopal to adapt to arid conditions even with very little soil. In dry conditions, the roots lose their absorbing hairs, and when water is available the hairs develop very fast and water and nutrient absorption becomes immediate.

The prickly pear is a uniocular, polyspermic, and fleshy berry shaped and sized as an egg. It is 4–12 cm in diameter and its color, depending on the species or varieties, may be soft green, or greenish white, canary yellow, lemon yellow, orange, red, cherry red or purple. The fruits have multiple lenticular-shaped, hard seeds distributed regularly all over the inside of the fruit. The seeds have a light testa, a wide aril, a curved embryo, large cotyledons, and a well-developed perisperm. One fruit may have up to 200 or more seeds which are impossible to chew. Rather, the whole fruit must be softly chewed, mixing the seeds with the pulp in the mouth to taste the sweet flavor of the juice and then swallowed wholly. Doing this for the first time can give you a very strange sensation. The characteristics of some commercial fruits are shown in table 1. The most marketed variety is the Reyna, but with time the ones with nice colors are becoming more popular. One important aspect is the existence of 'aborted' seeds in an amount three or more times higher than the fertile seeds. They are much smaller and also have pulp. Surely the fruits with more 'aborted' seeds may be the best from many points of view, as they are easier to ingest.

The nopal can be reproduced either by planting seeds which have a very slow starting growth (they require 90 days of development before being transplanted) or by asexual reproduction, planting cladodes or pieces of cladodes. In this case, the yield is better. The orientation of the cladodes related to sunlight is a major growth factor. It is required that the faces of the planted cladode be less exposed to direct sunlight to reduce evapotranspiration. This factor is of great importance in highlands where sun radiation is very intense.

Nutrient Content

The nopal is an excellent vegetable and fruit; it has many nutrients, in good quantities for some and small quantities for others. These have been of great advantage for the people living in the areas where nopal grows naturally and where other food availability is scarce. Some of the nutritional values of the edible parts and how they are eaten will be discussed below, including cladodes, prickly pears, and some processed products like the prickly pear cheese.

The average nutrient content per 100 g of edible portion of different parts of the plant is shown in table 2. The values are the average of many species [13]. As can be seen, cladodes are actually a very good vegetable. Hard fiber is a major component, with an average of 7% for six Opuntia species, as is shown in table 3 [14a]. The fruit has also a high fiber content, mostly from the indigestible seeds (table 4).

The prickly pear also compares favorably with other fruits. A recent study compared the most marketed variety, the Reyna, with apples and peaches. The whole fruit including the seeds [15] was analyzed (table 5).

Proteins and Fats

The protein content and amino acid composition of nopal cladodes are nutritionally significant. This vegetable is a very important food staple for people in the desert and semidesert areas in the Mid-Northern Region of Mexico. Protein

Table 2. Average nutritional values of 100 g of edible cladodes, prickly pear and prickly pear cheese of *Opuntia* spp.

Nutrients	Nopal (cladodes)	Prickly pear[1]	Prickly pear cheese[2]
Edible portion, %	0.78	0.55	1.00
Water content, %	90.1	91.0	17.2
Fiber, g	3.5	0.2	8.2
Energy, kcal	27	31	289
Carbohydrates, g	5.6	8.1	79.0
Proteins total, g	1.7	0.6	5.3
Lipids total, g	0.3	0.1	–
Cholesterol, mg	0.0	0.0	0.0
Calcium, mg	93	49	400
Iron, mg	1.6	2.6	13.0
Magnesium, mg	–	85	–
Sodium, mg	2	5	–
Potassium, mg	166	220	–
Retinol, eq, µg	260	5	8
Ascorbic acid, mg	8	22	88
Thiamin, mg	0.03	0.02	0.20
Riboflavin, mg	0.06	0.02	0.20
Niacin, mg	0.3	0.20	1.5

[1] Fruit without the seeds.
[2] Fruite milled with seeds and boiled to semi-dry.
From Muñoz de Chávez et al. [13].

Table 3. Hard fiber content of cladodes of some wild species of nopal

Scientific name	Wet, %	Dry, %
Opuntia ficus-indica	11.38	81.88
Opuntia amylacea	9.20	63.01
Opuntia megacantha	8.54	50.10
Opuntia robusta	5.72	46.93
Opuntia spp.	4.63	37.84
Opuntia streptacantha	2.73	18.57

From reference 14b.

Table 4. Fiber in 6 species of prickly pears (fruit with seeds)

Common name	Scientific name	Wet weight %	Dry weight %
Apastillada	*Opuntia ficus-indica*	11.38	81.88
Fafayuca	*Opuntia amyclaea*	9.20	63.01
Amarilla	*Opuntia megacantha*	8.54	50.10
Camuesa	*Opuntia robusta*	5.72	46.93
Verde	*Opuntia ficus-indica*	5.20	39.63
Cardona	*Opuntia streptacantha*	2.73	18.57

From Villarreal et al. [14c].

Table 5. Nutritional value of prickly pear (commercial variety) compared with apples and peaches

Nutrients	Prickly pear	Peach	Apple
Energy, kcal	38.0	46.0	65.0
Protein, g	0.6	0.9	0.3
Lipids, g	0.1	0.1	0.5
CHO, g	10.1	11.7	16.5
Fiber, g	2.4	2.4	2.0
Ascorbic, mg	31.0	19.0	11.0
Retinol eq., µg	4.0	22.0	3.0
Lysine, mg/100 g	4.0	3.7	4.0
Theonine, mg/100 g	4.8	3.4	3.1
Tryptophane, mg/100 g	0.8	0.4	0.1
Methionine, mg/100 g	0.7	3.8	1.7

concentrations in softer species, which are the preferred ones, are on average 0.6 g/100 g of wet weight. In harder varieties, the protein content can range from 1 to 2 g. The average figure in dry weight can be 12.2 g/100 g. Most importantly, the amino acid content of cladodes is quite valuable because it has significant quantities of lysine, methionine, and tryptophan [15], the amino acids in which corn and most grains are poor.

Table 6 shows the protein content of the prickly pear of different species and according to the place they grow. When only the digestible portion of the fruit is

Table 6. Protein content in some species of prickly pear

Common name	Scientific name	State	Water content, %	Proteins (N × 6.25), %
With seeds	*Opuntia* sp.	Reg. centro	77.3	1.31
White with seeds	*Opuntia hyptiacantha*	Hidalgo	86.2	–
Cardona without seeds	*Opuntia streptacantha*	Michoacán	91.0	0.55
Cardona with seeds	*Opuntia streptacantha*	Durango	88.1	1.06
Cascarona without seeds	*Opuntia hyptiacantha*	Durango	88.0	0.69
Chaveña without seeds	*Opuntia hyptiacantha*	Durango	90.8	0.55
Colorada with seeds	*Opuntia robusta*	Hidalgo	82.0	–
Mansa without seeds	*Opuntia ficus-indica*	Michoacán	89.4	0.93
Mansa white with seeds	*Opuntia ficus-indica*	Hidalgo	87.8	1.21
Mansa colorada with seeds	*Opuntia robusta*	Hidalgo	85.7	1.43
Tapona with seeds	*Opuntia robusta*	Hidalgo	86.3	1.93
Redonda with seeds	*Opuntia* sp.	Hidalgo	85.6	1.62
Red with seeds	*Opuntia streptacantha*	Hidalgo	82.6	1.81

From reference 14b.

analyzed, i.e. the pitted prickly pear, protein values are low, as the digestible material is mostly formed by water and soluble carbohydrates. The seeds are not digestible, but can be utilized when the fruit is milled and cooked to form prickly pear cheese which, due to its concentration, has an important protein content.

As all vegetables, nopal has a low fat content, although some varieties can have up to 3% in dry weight, which can be significant, especially since preliminary analyses show that a high proportion of the fatty acids may occur as ω3 [13].

Vitamins

Table 2 shows that in cladodes the amount of β-carotene (expressed as µg of retinol equivalents), is high, 260 µg/100 g, which is especially important considering the size of the servings consumed and that in semidesert populations no other source of vitamin A is easily available [13a]. During the harvest season, the consumption of prickly pears per person can be huge, at least ten prickly pears per person per day. The fruit has many other carotenoids not yet fully identified at this time. The variety in colors and their intensity suggests the possibility of their presence, although it is not certain whether they can metabolically produce vitamin A or not.

Ascorbic acid content varies with species, but given the amounts consumed, the daily allowances of vitamin C can be met with the daily normal consumption

Table 7. Mineral content in some species of nopal cladodes

Sample	SiO$_2$	CaO	K$_2$O	MnO	MgO	N$_2$O	CO$_2$=	SO$_4$=	Cl–	P$_2$O$_5$
Opuntia ficus-indica	1.60	42.55	11.06	0.12	6.05	1.28	39.08	2.95	2.84	1.25
Opuntia hyptiacantha	1.82	42.65	11.01	0.28	7.89	0.26	34.23	1.54	2.15	1.50
Opuntia tormentosa	3.79	45.65	10.12	0.26	6.73	0.81	34.50	1.29	0.95	1.05
Idem, young cladodes	3.32	44.16	8.87	0.18	5.88	0.47	38.12	1.16	0.93	1.97
Opuntia megacantha	1.24	46.35	8.13	0.20	5.92	0.36	38.63	1.03	0.99	1.26
Opuntia robusta	1.18	46.35	9.63	0.25	6.15	0.83	33.60	1.27	2.02	1.38

From reference 14b.

of cladodes, or 1–2 prickly pears. This is quite good, especially when considering that nopal may be the only source of vitamin C in many dry areas.

As to other vitamins, such as thiamine, riboflavin and niacin, the contents are low, both in the cladodes and the fruit.

Minerals

Mineral content varies a great deal, both quantitatively and qualitatively, not only between the various species but also within the species. These variations are the result of genetic and climatic conditions, and perhaps also of soil composition.

Calcium and potassium are the main minerals found in nopal. Its potassium content is of major importance to the diet in the nopalera regions, since bananas or oranges, or other good sources of minerals are virtually unavailable. The nopal also contains magnesium, silica, sodium, small quantities of iron, aluminum, calcium and manganese in the form of carbonates or chlorides, sulfates and phosphates (table 7).

Carbohydrates

Carbohydrates are the most abundant components in the nopal vegetable, as the plant synthesizes a large variety of hexoses, pentoses and their polymers as cellulose, hemicellulose and many kinds of mucilages. The high content of total fiber in nopal, and especially of the so-called soluble fiber, is of enormous significance because it may be the source of some of the effects attributed to its consumption like decreasing hyperglycemia, lowering cholesterol, and improving gastric acidity (table 8).

Table 8. Types of carbohydrates in nopal cladodes (sample of *Opuntia tormentosa* and *Opuntia robusta*)

Carbohydrates	Percent of wet weight
Total sugars	10.41
Total polysaccharides	8.49
Total disaccharides	1.60
Total monosaccharides	0.32
Total hexoses	3.78
Hexose polysaccharides	1.97
Hexose disaccharides	1.55
Hexose monosaccharides	0.26
Total pentosanes	5.12
Pentose monosaccharides	0.10
Other pentosanes	5.02
Other CHO (uronic, etc.)	1.70

From Fernandez Landero [13b].

Organoleptic Qualities

In addition to the nutritional values of the nopal, its different parts also have sensory qualities like the delicious sweet flavor of the fruit and the pleasant texture and mild flavor of the cladodes, making them easy to combine with other foods. There are many regional recipes on their use as a vegetable, which always is consumed boiled or fried. The most common recipes with cladodes include a small piece of meat, a boiled egg, or some dry fish (during Lent), cooked with chili sauce, traditionally always consumed with beans and tortillas. The cooking qualities of the nopal place it in a prominent position in the Mexican cuisine.

In summary, it can be said that both the prickly pear and the cladodes, as well as some of the byproducts such as the prickly pear cheese, are very nutritious, especially considering the frequency and size of the usual servings. The plant contributes a large amount of carbohydrates, proteins, minerals, vitamins, and even water. These nutrients, no doubt, play a significant nutritional role in poor regions with low agricultural and livestock production. For large city dwellers, with good nutrient intake, nopal has the qualities of a good vegetable, rich in fiber, and perhaps with some additional benefits for the prevention or even the treatment of certain metabolic disorders related to chronic diseases and obesity, as the cladodes have the advantage of being a low-calorie food.

Other Health and Nutrition Characteristics

It is an old Mexican tradition to use nopal as a home remedy. References on its medicinal properties are found in the Conquest Chronicles [1, 2]. Not surprisingly there is growing interest in studying the medical uses of nopal.

In talking with farmers of the Mexican high plateau, one has the impression that time has stood still. They continue using nopal for its many medicinal properties, both by eating it or using it topically as poultices for people and animals. The varieties most widely used for curative purposes are the giant cactus (*O. streptacantha* Lem.), the xoconostle (*Opuntia joconostle* Weber), the nopalillo (*O. macrorhiza* Englem.), and the teasel *(Opuntia imbricata)* [12].

Mexican traditional medicine includes the empirical use of different preparations made with the nopal cladodes mainly to treat gastritis and peptic ulcers, fatigue and dyspnea. In other regions, it is used to improve glaucoma and capillary fragility. The cladodes cut in small pieces and boiled are used to treat various liver conditions, like 'liver congestion and hypertrophy', resulting from alcohol abuse.

People use warm cladodes applied locally for rheumatic pain, on fresh wounds, burns, and chronic skin ulcers.

Besides those traditional uses of nopal, some new applications with more scientific base are now being tried: mostly the consumption of nopal cladodes for the prevention or treatment of hyperglycemia, gastric acidity and atherosclerosis. The most wide spread use throughout Mexico is for the treatment of diabetes [23]. According to popular information, diabetics improve their symptoms by ingesting nopal cladodes for long periods of time.

Aqueous and dry cladode extracts have been tried in various studies on experimental animals (dogs, rats, rabbits, guinea pigs), both normal and with experimentally induced hyperglycemia. All the studies show that even though its effect is not significant in normal animals, it is constant although not very pronounced in the diabetic and hyperglycemic ones [16]. Some studies have been conducted to evaluate the different chemical compounds of the nopal, like the tannins, the reducing and non-reducing sugars, the flavonoids, and some of its alkaloids, but none of these components seems to be responsible for the hypoglycemic effect found in the complete cladode [17].

The information on humans, obtained by experimental studies in healthy and diabetic individuals [18–22] may be summarized as follows:

(1) In healthy individuals, nopal intake does not significantly modify fasting glucose levels and insulin serum concentrations.

(2) In healthy individuals, nopal intake reduces the elevation of glucose and insulin serum concentrations occurring after an oral glucose load.

(3) In individuals with non-insulin-dependent diabetes mellitus, nopal intake significantly reduces glucose and insulin fasting serum concentrations.

(4) In individuals with insulin-dependent diabetes mellitus, nopal intake before usual meals for 10 days significantly reduced the glucose serum concentrations.

In a study conducted in healthy volunteers, glycemia and insulin release were reduced when nopal was used, while glucagon, cortisol and growth hormone levels showed no differences. Some data show that insulin antagonist hormones are not involved in the hypoglycemic action of the nopal [23].

Studies conducted at the National Nutrition Institute, suggest another possible hypoglycemic mechanism, probably secondary to a mechanical effect of gastric distension and the release of some enterohormone. This results in greater insulin secretion during the early stage and, consequently, in reduced glucose and insulin levels during the late stage [24].

Equally or more important than the above are the beneficial effects of nopal on circulating cholesterol, as suggested by some publications [25]. A study on healthy and diabetic subjects, found that nopal intake before every meal for 10 days reduced body weight and total cholesterol, triglyceride, and LDL cholesterol concentrations. In many of the studies a placebo effect cannot be excluded; therefore this beneficial effect is not definitively proven.

The possible mechanisms of action of its metabolic effects are also not well defined, i.e. the existence of active hypoglycemic or hypolipidemic principles is not known. However, this may be due to the high content of several soluble fibers (pectins, mucilages). The results of long-term nopal treatments of hyperglycemic or hypercholesterolemic subjects are still unknown, because none of the mentioned reports is conclusive, due to the fact that their experimental designs leave much to be desired. It is badly needed to have better designed studies, for example using another vegetable as control.

The effect of prickly pear intake on LDL in guinea pigs with induced hypercholesterolemia has also been studied. It reduced LDL cholesterol plasma and liver concentrations and increased liver B/E apolipoprotein receptors. As in the case of the vegetable part of the plant, it is possible that it acts as a mechanical bile-acid-sequestering factor [26].

Studies with rats conducted at the Institute of Nutrition show that the nopal extract has a pH-buffering effect that protects the gastric mucosa from certain aggressive agents such as aspirin. This suggests nopal might be useful in treating peptic acid disease. Some well-conducted clinical trials are needed [27].

In summary it can be said that it is necessary to carry out in-depth clinical studies to ascertain the true effect and mechanism of action of the nopal for the prevention and treatment of the chronic diseases mentioned, and the mechanisms of its various effects. For the time being, the consistency between the studies in

Table 9. Estimated changes in the vegetable nopal market in major cities in thousand tons

Years	Demand	Supply
1977	211	120.0
1978	216	124.6
1979	219	129.3
1980	225	134.2
1981	247	139.3
1982	250	144.6
1983	263	150.1
1984	275	155.8
1985	280	161.7
1986	283	167.8
1987	304	174.2

From reference 28a.

laboratory animals and the few clinical trials is very notable and some credit should be given to the nopal as a supportive measure. This means that it may be used as an adjuvant in the treatment of chronic diseases.

Actually, the use of nopal in all of its forms as an antidiabetic agent is becoming a widespread practice in Mexico. In countries such as Japan and Germany, nopal is already imported for this purpose. Perhaps in the future nopal will be advised as one of the least aggressive and most economical options for controlling the mild forms of diabetic and hyperlipidemic conditions. There is an old saying 'let your food be your medicine', and this may be a true statement for nopal in relation to some chronic noncommunicable diseases.

Production and Consumption

Production

Unfortunately, data on production and demand have not been systematically recorded. In 1976 in the main cities of Mexico the supply of nopal vegetable was estimated at 97,204 tons; 11 years later, the figure published was 174,200 tons (table 9). In general, the production reports show a gradual increase in growing surfaces and in the efficacy of new production techniques. A recent significant figure (1990) shows that about 40,000 tons were sold on the sole Central Market of Mexico City.

Adequate control of cropping, fertilizers and weeding allows for intensive production of the vegetable all year round with maximum yields of up to 75 tons per hectare per year [28b].

The highest vegetable production of nopal occurs immediately after the first rains but at that time, because of oversupply, prices drop, usually resulting in economic losses. This happens during the dry hot season, between March and June, and most producers do not crop all of the nopal they can due to the low pricing in the market. This also applies to semi-wild nopaleras, whose only cropping time coincides with the high productivity of cultivated nopaleras causing a serious economic problem for them. Conversely, in winter, production is very low or virtually nil in places where no irrigation is available, as in the case of major wild growing areas.

Horticulture. Nopal grown as a vegetable is optimally planted at a 40,000 plant per hectare density, separated 30 cm from each other from cladode center to center, with a 10 cm clearance between rackets or plants and 80 cm between furrows. In this case organic fertilizers are used and young or tender cladodes are cropped when they are 2 or 3 months old. These nopal buds are eaten as vegetables and the demand for them is high especially between February and April.

The cladodes have an average shelf life of 28 days under controlled conditions at 10°C and an 80–85% relative humidity.

The market for the nopal as a vegetable is constantly growing: at present there is a very high demand in Japan, the US and some European countries. There are two small industries packing pickled nopal, a product widely accepted both domestically and abroad. An alternative industry is the supply of dehydrated nopal in capsules that are being sold mainly in health food stores and drug stores. Because of the high demand for this product, the number of industries currently engaged in dehydrating nopal tend to increase. Today, nopal is sold in capsules, powder and tablets for popular, empirical, non-prescription supplement, with the belief that nopal has pharmacologic properties previously described.

Fruit Growing. Prickly pears have shown a noticable market growth during the last few years, mainly in major cities that are showing very large consumption capacity. The prickly pear has also become popular in high class social circles and is found on the menus of large hotel chains and in fine restaurants, but its main market is the working class. A typical scene in Mexico City is a street stand with peeled prickly pears on a stick, placed on ice blocks for sale. And you can order them spiced with chile powder, salt or lime.

The prickly pear has a relatively long life span and can be eaten fresh 3–4 weeks after harvest. Not much work has been conducted to extend this preservation period. It is industrialized on a regional basis, and almost exclusively in traditional ways, i.e. as prickly pear cheese. Lately, some new attempts such as prickly pear wine have been tested. Other technologies have been tried, but with no economic results. To encourage new technologies, both in production and industrialization, organized producer groups and agricultural universities are holding meetings every other year to submit and discuss these important issues. Beginning in 1990, these meetings have become international and gather specialists who face similar problems in other parts of the world, mostly in Italy and Chile [29].

Fodder. The use of the nopal as fodder is very important in Mexico, especially in livestock growing areas in the Nothern regions. The plant, because of its perfect adaptability to arid conditions, is virtually the only food available for ruminants of domesticated species, mostly cattle and goats. Nopal is used, after taking out the thorns, as a source of water and auxiliary fodder when mixed with molasses during droughts or when food is scarce. Goats and sheep livestock can eat nopal almost all year round [30].

The nopal species most widely used as fodder are *Opuntia cantabrigensis, Opuntia lindheimeri, Opuntia leuchotrichia, Opuntia estreptacantha, Opuntia rastrera, Opuntia microdasys, Opuntia pilifera, Opuntia maxima,* and *Opuntia robusta.* All of these species grow wild. Plantations for fodder are very scarce although many studies have shown that nopal grown for fodder can be very successful economically [31].

Raw protein and general nutrient content of the fodder varieties vary greatly from species to species. When fresh, protein content ranges from 2.8% in *O. rastrera* to 8.8% in *Opuntia stenopetala.* Nopal is also a source of major quantities of minerals such as calcium, potassium, phosphorus, sodium, fiber, and some vitamins, that are all essential nutrients for livestock. Some authors have indicated that Holstein cows, when supplemented with a concentrate, eat around 57.0 kg of nopal per day. When no concentrate is available, the average daily nopal consumption by milk cows can range from 77 kg to as much as 117 kg [31].

Scarlet Dye Production. Scarlet dye is produced from cochineal insects *(Dactylopius cocus costa)* which develop in 2-year-old or 3-year-old nopaleras whose cladodes and fruits have the required surface for cochineal eggs to be placed. At the end of 3 months, cochineal insects are fully developed and ready for cropping. About 140,000 insects are required to produce 1 kg of scarlet dye. Cochineal insects also provide the carminic acid used in the food, pharmaceutical, and textile industries [32].

With new production systems, yields can be as much as 2,840 tons of cochineal insect per hectare of nopal planted. Pricing for the carminic acid ranges from 66 to 100 US$ for 1 g so its production can be quite profitable. The main producing state in Mexico today is the Southern state of Oaxaca. The National Polytechnic Institute (IPN) is training farmers on production techniques. At present, nearly 90% of this Mexican product is exported to the United States of America which also imports it from Peru and the Canary Islands in Spain [33].

Consumption

Nopal is consumed as a vegetable almost all year round in cities and in areas where it is grown. Consumption of prickly pears, on the other hand, is much more limited because it is consumed fresh during the rainy season. Although limited to its short growing season it is abundant in quantity. In dry, hot areas its demand is quite substantial, almost 1 kg per person per day.

Regarding the high seasonal availability of prickly pears, an interesting hypothesis was proposed on the possible relation of its excessive temporary consumption with certain metabolic conditions of certain Indian groups of the semi-arid North plains. Maybe for hunter-gatherers their ability to obtain energy from the prickly pear, store it as fat and later to mobilize it as glucose for hunting, was a surviving genetic trait, which later, when sedentary, made them more susceptible to diabetes. Therefore, the high prickly pear production in a short season contributed to the genetic selection of individuals with the capacity to store much energy during the intensive collection times, thus helping them to survive in times of shortages. This metabolic adaptation of the native groups could be at the origin of an important health risk for Mexicans today who tend to become obese and diabetic as a result of both a sedentary life style and consumption of large quantities of food.

Data obtained from the National Rural Dietary Food Survey, conducted by the National Nutrition Institute in 1989 in 19 areas in Mexico are included in table 10. The survey found low consumption rates [34]. The data were reanalyzed for this paper and some recording failures were found. The main source of error came from the fact that many people are ashamed to admit they eat nopal, as it is traditionally consumed by 'the poor' and therefore has low prestige, as is the case with consumption of other native vegetables such as purslane and chayote. In future studies, surveyors should insist in asking specifically about nopal consumption, as the present questionnaire asks on the consumption of vegetables and fruits in general, and curiously nopal is not considered a vegetable as the rest. As it is actually a staple food, the questionnaires should mention it specifically.

Table 10 also shows that the small proportion of the population that declared eating nopal declared, as expected, consuming really large amounts.

Table 10. Daily family consumption of nopal as vegetable by regions national household survey INNSZ 1989

Zone	Families surveyed	Families that consumed %	Average intake in total families g/day	Intake in families that consumed g/day
1	438	3.88	10.7	629.4
2	328	2.44	6.3	787.5
3	571	2.10	9.6	802.5
4	282	3.19	4.6	508.9
5	739	4.60	21.2	622.9
6	281	4.98	6.1	433.6
7	714	4.62	15.4	467.5
8	436	15.37	36.5	544.1
9	1,172	8.70	48.5	475.1
10	1,703	8.46	90.4	627.7
11	1,371	1.60	12.4	565.0
12	1,130	16.02	119.2	658.4
13	1,191	1.18	6.5	461.4
14	737	4.48	19.8	600.9
15	1,090	0.18	1.1	540.0
16	1,645	3.89	39.8	621.4
17	744	0.13	0.2	160.0
18	735	0.51	0.0	–
19	812	0.12	1.5	1,500.0
Total	16,119	0.59	449.6	593.1

From reference 34.

The range of different dishes that can be prepared with the cladodes is impressive. There are over 100 popular recipes in which nopal is combined with traditional foods and there are many other highly sophisticated recipes that are served in fine restaurants.

Fresh prickly pears served on ice is still the preferred choice. However, during the growing season, prickly pear juice is served in almost all coffee shops, hotels, and restaurants, and even in bars where prickly pear juice is used as a cocktail base. Prickly pear cheese and 'colonche', a fermented drink made of prickly pear juice, are the next most popular choices, although these are limited to the Central and Mid-North regions of the country.

Conclusions

The nopal is a noble and generous plant, adapted to arid and semiarid areas where it grows wild. It can, however, be commercially grown with very good results. It adapts to droughts and high temperatures and possesses a metabolism which can be four to five times more efficient in converting water to dry matter than the most efficient grasses. It is one of the most remarkable plants in the world because of its appearance and the capability of being used totally, including the stem, the flowers, and fruits to please the palate, as fodder for stable or grazing animals, as a source of fuel, to glue adobe bricks in rural homes, as medicine, in ritual ceremonies, to embellish the countryside, as a host of a foreign insect producing a beautiful natural dye, and as a source of many nutrients, including some very important antioxidants.

In spite of its abundance, excellent qualities, domestic demand, and potential for exports, it is a 'neglected' crop on which there are virtually no agricultural statistics or plans to encourage its production. Records on consumption are also inadequate.

More studies are required on biochemical and molecular, ethno-botanic and phytogenetic aspects, and more research is needed on its special potential regarding the prevention and treatment of chronic-degenerative diseases. Above all, it is important to perform better human research on the blood cholesterol and blood glucose lowering effects of nopal.

Finally, we wish to emphasize the need for more worldwide information on the uses and bounties of nopal, considered monstrous by some, beautiful by others, and magic for many, although for Mexicans, throughout their history, it has meant food, delights, medicine, and the symbol of their homeland.

References

1   Sahagún B: Historia general de las cosas de la Nueva España. México, Porrúa Hnos, 1987.
2   Clavijero FJ: Historia antigua de México. México, Porrúa Hnos, 1986.
3   Badiano J: Libellus de Medicinalibus Indorium Hervis. Aztec manuscript, México, 1552.
4   Brom Rojas F: El Nopal. Comisión Nacional de Fruticultura. Secretaría de Agricultura y Ganadería, México, 1970.
5   Bravo-Hollins H: Las Cactáceas de México. Universidad Nacional Autónoma de México, 1978, vol 1, pp 6, 167.
6   Halhali Baghdad A: Personal communication. National Institute of Nutrition, Mexico, 1993.
7   López G, Elizondo JJ, Elizondo JL: El conocimiento y aprovechamiento del nopal en México. Conferencia inaugural de la Tercera Reunión Nacional y Primera Internacional sobre el Nopal: Su Conocimiento y Aprovechamiento, Saltillo, 1990.
8   De Kock GC: Manejo y utilización del nopal sin espinas. 9°. Anales Congreso Internacional de Pasturas, 1965, vol 2, p 1471.
9   Salgado Molina C: El Cultivo del Nopal, Una Alternativa Económica en Suelos Aridos y Semiáridos, CODAGEM, Secretaria de Agricultura y Recursos Hidráulicos, México, 1984.

10 Muñoz de Chávez M: Reporte de la misión de IFAD sobre la situación alimentaria y nutricional de la población de la zona ixtlera. IFAD/FAO Rep 427-90 Roma, 1990.

11 Piña-Luján M: Observaciones sobre le grana y sus nopales hospederos en el Perú. Rev Soc Mex Cact Sucul 1976;26:10–15.

12 Waizel B, Pulido A: Communicación acerca del uso de *Opuntia dillenis* en homeopatia. Memorias de la Reunión Nacional sobre el Conocimiento y Aprovechamiento del Nopal, México, 1987.

13a Muñoz de Chávez M, Hernández M, Roldán JA: El valor nutritivo de los alimentos de mayor consumo en México: Tablas de uso práctico. Mexico, Solidaridad-INNSZ, 1992.

13b Fernández Landero A: Estudio químico de seis muestras de Nopal del Valle de México (1949); in Bravo-Hollins H: Las Cactáceas en México. Ed Universidad Autónoma de México, 1978, Tomo I, pp 62–83.

14a Bauer R, Flores VC: Análisis bromatológico de cuatro variedades de *Opuntia ficus-indica* en Chapingo. México, Escuela Nacional de Agricultura, 1969, monogr No 27.

14b Instituto Nacional de Investigaciones Forestales (INIF) – Comision Nacional de Zonas Aridas (CONAZA): El Nopal. Mexico, INIF-CONAZA, 1981, p 63.

14c Villarreal F, Rojas-Mendoza P, Arellano V, Moreno J: Estudio químico sobre seis especies de nopales (Opuntia Spp). Ciencia, México, 1963;22(3):59–65.

15 Mendoza E: Personal communication. Mexico, Laboratory of the National Institute of Nutrition, 1993.

16 Frati-Munari AC, Fernández-Harp JA, Quiroz-Lázaro JL: Acción hipoglucemiante de diferentes especies de nopal *Opuntia streptacantha* Lemaire en pacientes con diabetes mellitus tipo II. Arch Invest Méd México 1989;20:197.

17 Frati-Munari AC, Quiroz-Lázaro JL, Altamirano-Bustamante P, Ariza-Andraca CR: Efectos de diferentes dosis de nopal *Opuntia streptacantha* Lemaire en la prueba de tolerancia a la glucosa en individuos sanos. Arch Invest Méd México 1988;19:143–148.

18 Frati-Munari AC, Gordillo BE, Altamirano-Bustamante P, Ariza-Andraca CR: Disminución de glucosa e insulina sanguínea por nopal *Opuntia* spp. Arch Invest Méd México 1983;14:269.

19 Frati-Munari AC, Quiroz-Lázaro JL, Altamirano-Bustamante P, Banale-Ham M, Islas-Andrade S, Ariza-Andraca CR: Efecto de diferentes dosis de nopal (*Opuntia streptacantha* Lemaire) en la prueba de tolerancia a la glucosa en individuos sanos. Arch Invest Med 1988;18:234.

20 Frati-Munari AC, Gordillo BE, Altamirano-Altamirano P, Ariza-Andraca CR: Hypoglycemic effect of *Opuntia streptacantha* Lemaire in NIDDM. Diabetes Care 1988;11:63.

21 Frati-Munari AC, Fernández-Harp JA, Banales-Ham M, Ariza-Andraca CR: Disminución de glucosa e insulinas sanguíneas por nopal (*Opuntia* sp). Arch Invest Med 1983;14:269.

22 Frati-Munari AC, Fernández Harp JA, Torres MC: Influencia de un extracto deshidratado de nopal (*Opuntia ficus-indica* mill) en la glisemia. Arch Invest Méd 1989;20:212–216.

23 Fernández-Harp JA, Frati-Munari AC, Chávez-Negrete A, De la Riva H, Marez-Gómez G: Estudios hormonales en la acción del nopal sobre la prueba de tolerancia a la glucosa. Informe preliminar. Rev Med IMSS 1984;22:387.

24 Bustamante JF, Rivera R, Wong-Chaires B, Gómez-Pérez FJ, Rull JA: Efectos del nopal (*Opuntia* sp.) sobre la fase rápida de secreción de insulina y la glucemia en individuos sanos. En Memorias de la XXVIII Reunión Anual de la Soc Mex de Nut y Endocrinol, México, 1988, p 72.

25 Frati-Munari AC, Fernández-Harp JA, Torres MC: Efecto del nopal (Opuntia sp.) sobre lípidos séricos, la glucemia y el peso corporal. Arch Invest Med 1983;14:117.

26 Fernández ML, Trejo A, Mc Namara C: Pectin isolated from prickly pear (Opuntia sp.) modifies low density lipoprotein metabolism in cholesterol in guinea pigs. Publication of the American Institute of Nutrition, New York, N.Y., 1990, p 1283.

27 Gulias A, Gálvez E, Robles G: El nopal como amortiguador de la acidez. Rev Invest Clín 1989;41: 387.

28a A scientific status summary by the Institute and Food Technologist's expert panel food safety and nutrition. Food Technol 1979;January:35–39.

28b Anuario Estadístico (Statistics Yearbook). Secretaría de Recursos Hidráulicos. Dirección General de Estudios Agrícolas. México 1988.

29 López GJJ, Ayala OMJ: El nopal: Su conocimiento y aprovechamiento. Memorias Primera Reunión Internacional, Universidad Autónoma 'Antonio Narro', Saltillo, 1990.

30 Flores Valdez C, Aguirre Rivera JR: El nopal como forraje, Universidad Autónoma de Chapingo, México, 1979.

31 Belasco IJ: The response of rumen microorganisms to pasture grass and prickly-pear cactus following foliar application and urea. Anim Sci 1958;17:209.

32 Piña-Luján C: La grana o cochinilla del nopal. México, LANFI, 1977, monogr No 1, p 46.

33 Instituto de Investigación Tecnológica Industrial y de Normas Técnicas (ITINTEC): Manual Técnico Sobre la Tuna y la Cochinilla. ITINTEC-CIID, México, 1991.

34 Instituto Nacional de la Nutrición Salvador Zubirán: Encuesta Nacional de Alimentación en el Medio Rural. México, INNSZ, 1990, monogr No L-87.

Miriam Muñoz de Chávez, Instituto Nacional de Nutrición S.Z., Community Nutrition Division, Vasco de Quiroga 15, Tlalpan, México, D.F. CP 14000 (México)

Simopoulos AP (ed): Plants in Human Nutrition.
World Rev Nutr Diet. Basel, Karger, 1995, vol 77, pp 135–146

....

The Corn Tree *(Brosimum alicastrum)*: A Food Source for the Tropics

M. Ortiz, V. Azañón, M. Melgar, L. Elias

Institute of Nutrition of Central America and Panama,
Pan American Health Organization, Guatemala City, Guatemala

Contents

Introduction 135
Botanical Description 136
    Morphology 137
Uses 138
    Food 138
    Fodder 138
    Wood 139
    Medicinal 139
    Other Uses 139
Distribution 139
Production 139
Nutritive Value 140
    Chemical Composition 141
    Amino Acid Content and Protein Quality of the Corn Tree Seed 142
Development Potential 143
    Tortillas and Bread 143
    Animal Feed 144
    Forestal Resources 144
Conclusions 145
References 145

Introduction

Corn and *Brosimum alicastrum* were probably the main food sources for the Mayas of the classical period, one of the most advanced ancient civilizations of the American continent [1]. According to Puleston [2], the Mayas cultivated *B. alicastrum* intensively and used the seed as part of their diet. The Kaqchikel (Guatemalan highland indians) gave *B. alicastrum* ceremonial significance, using the name 'iximchee'; this can be literally translated as 'corn tree' (ixim = maize, chee = tree) (fig. 1). They even called their most important city Iximchee [2, 3].

In 1937, Lundell [4] reported finding many of these trees near the sites of Mayan ruins. Later, Puleston [2] established the association of *B. alicastrum* with both ceremonial and dwelling areas in research carried out at Tikal, Petén. This supports the theory that the *Brosimum alicastrum* trees currently found in dwelling areas are descended from trees grown by Mayas near their houses in a sort of garden plot, and are probably a relic of ancient horticulture. However, Lambert [5] sustains the thesis that the association between *B. alicastrum* and dwelling areas is not socioeconomic, but purely ecological, since the dwelling areas have favorable conditions for the development of this species. Currently, *B. alicastrum* is still used in the human diet in certain Guatemalan communities of Maya descent [6].

In 1975, the US National Academy of Sciences [7] included *B. alicastrum* among a group of 36 plants considered to be underexploited, while having a high economic potential for tropical regions.

Botanical Description

In nature, this large evergreen tree may attain a height of 40 m and a diameter of more than 1 m [8].

Taxonomy

Kingdom	=	Vegetable
Subkingdom	=	Embryobionta
Division	=	Magnoliophyta
Class	=	Manoliopsida
Subclass	=	Hamamelidae
Order	=	Urticales
Family	=	Moraceae
Genus	=	Brosimum
Species	=	approximately 28

Standley and Steyermarck reported four species in Guatemala [3, 6]; *B. alicastrum* Swartz, *B. panamense* Pittier, *B. costarricanum* Liemb, *B. terrabanum* Pittier.

Fig. 1. The corn tree. Tikal, Guatemala.
Fig. 2. The trunk. Tikal, Guatemala.

According to Aragón, cited by Asenjo [3], 'Brosimum' is derived from the Greek word 'Brosimos', which means edible. It has many common names, such as ramón, ujuxte, ojoche, iximché, capomo, másico, ox [1, 3, 6, 9], of which ramón is the most popular in Guatemala. For purposes of clarity, the name corn tree will be used hereafter.

Morphology
Trunk: A medium-sized to large tree, measuring 18–45 m (fig. 2).

Bark: The inner bark is fibrous, creamy-yellow with large amounts of latex. The outer bark is smooth or scaly, and dark or light gray in color [3, 6, 9].

Branches and leaves: First ascending then drooping; leaves are sometimes oval-shaped with scars from fallen stipules, gray-green or glabrous.

Flowers: Male inflorescences have abundant cream-colored flowers, either with a barely visible perianth or without (fig. 3). Each flower has a single stamen with a peltate anther, circumscissily dehiscent and surrounded by peltate bracts. Female inflorescences are smaller and have a long style with two stigmata [9].

Fruit: Yellowish-green during maturation and orange when ripe (fig. 4). Fruits are subglobose, and 2–2.5 cm in diameter. They have a fleshy pericarp and the surface is covered with many white scales, from which one seed is obtained [3, 6, 9]. The pulp is edible and sweet. Fruiting occurs from December to July, for a period of 50–75 days. One kilogram of fruit consists on average of 189 units; the fresh seed weight accounts for 47%, and the pericarp for 53%. The seed accounts for 80% of the weight of the dry fruit [6].

Seeds: One 1.2-cm seed is obtained from each fruit (fig. 5). It is covered by a yellowish papery testa, with two asymmetrical greenish (when fresh) cotyledons. When cut, they yield large amounts of latex [3, 6].

Latex: Abundant, white, with low viscosity; slightly salty taste, it congeals slowly [6].

3

5

4

Fig. 3. Corn tree flower. Tikal, Guatemala.

Fig. 4. Corn tree fruit and leaves.

Fig. 5. Corn tree seeds. Tikal, Guatemala.

Uses

Food

The corn tree is one of the tropical species of which every part may be used [3]. The fruit's pulp is edible, and may be eaten raw, boiled or made into juice or marmalade [8]. The seed has a flavor similar to that of the potato. It is eaten roasted (like chestnuts) or boiled and used as a potato substitute. When the seeds are roasted and milled, they are steeped in boiling water to make a beverage, or used as flour; either alone or mixed with corn to make different dishes such as tortillas [9]. According to reports, since the Mayan classical period, and until now, the seed has in fact replaced corn when this grain has not been available [3, 6].

Fodder

The most common use for this plant is as fodder, becoming more important in dry areas, since this tree remains green during the dry season, and has many leaves, digestible stalks and seeds. Thus, it is prized as forage for cattle. The practice of cutting fodder from trees that grow wild is not efficient; however, on plantations with dense, technologically handled areas of tree cultivation, cattle may even browse among the trees, as they are easily reached. The leaves have an average crude protein content of 30% of dry matter base [10].

When comparing the corn tree with conventional fodder sources, it was found that it has high productivity in quality and quantity [10]. Other studies developed on the seed have shown that it may partially replace sorghum in feed for poultry and pigs [11, 12].

Wood
Since corn tree wood is soft compared to other tropical trees, it is easy to handle and widely used in making inexpensive furniture and other utensils [3].

Medicinal
Corn tree fruit, leaves, latex and bark have traditional therapeutic uses among the population both in Mexico [8] and Guatemala [6]. The seed is used as a galactogogue, the latex to fight cavities, the leaves for cough syrup and the bark for asthma [6, 8]. At this time no formal evaluation of these health claims has been carried out.

Other Uses
The tree is also used as an ornamental plant, and to provide shade in coffee plantations. It is considered very important as a contributor to environmental protection [6].

Distribution

The corn tree is found in humid and very humid hot subtropical forests, at altitudes from 300 to 2,000 m above sea level. Its most common habitat is at 125–800 m above sea level, with a yearly rainfall of 1,500 to 2,000 mm and average temperatures of 24.4°C; relative humidity from 80 to 96%, and 30 to 50% daylight [6].

The corn tree may be found in both northern and southern Guatemala, the southern part of Mexico, Belize, El Salvador, the Caribbean, and Hawaii [1, 6, 8].

Production

Estimates for production are as follows: an adult tree (more than 8 years old) in a high forest may yield up to 75 kg of fruit, 58 kg of seed, and 400 kg of green leaves [8, 9]. In densely wooded areas, 250 trees/ha may be estimated. A planta-

Table 1. Proximal chemical composition of the seeds of *B. alicastrum* (percent)

Type of seed	Reference source	Water content	Ash	Ether extract	Crude protein	Crude fiber	Free nitrogen extract
Seeds Tamaulipas (Mexico)	3	47.96	2.08	1.06	6.83	2.42	39.65
Dry seeds Veracruz (Mexico)	3	12.17	4.21	2.02	10.22	8.90	62.48
Dry seeds Veracruz (Mexico)	3	4.60	3.84	0.86	9.95	–	–
Seeds	3	6.05	4.4	1.6	11.4	6.2	69.9
Green seeds Petén (Guatemala)	1	6.22	4.20	2.52	8.9	3.94	74.22
Ripe seeds Petén (Guatemala)	1	5.91	4.11	2.76	8.6	4.02	74.60
Gather seed Petén (Guatemala)	1	9.38	3.94	2.43	7.7	4.00	72.55

tion of 100–125 trees/ha may allow planting in association with other plants to be used for food or other crops of agro-socioeconomic value.

In a 3-year study of corn tree seed yield per hectare in natural forests in Tikal (Guatemala), an average of 1,762 kg with a maximum of 2,616 kg of seed per year was found [1, 3, 6, 9].

Nutritive Value

Since the 1960s, efforts have been made to evaluate the nutritive potential of the corn tree seed. A wealth of information may be found in the available literature regarding the chemical composition [6] of the corn tree seed, slightly less on the seed's amino acid pattern [9] and few data from biological assays performed to evaluate nutritional quality [3].

Analysis of the fruit pulp yields 84% water content, 2.5% protein, 0.5% ether extract, 1.2% fiber, 10.9% free nitrogen extract [13].

Since the seed has been used most, more studies are available on this topic [3, 6, 8, 9]. Fresh seeds may have as much as 52.2% of water content. This makes it necessary to carry out a dehydrating process to store them. After drying, the water

Table 2. Chemical composition of the corn tree fruit (whole fruit)

Item	Composition, %
Water content	8.0
Crude protein	12.3
Crude fiber	27.3
Ether extract	3.1
Ash	15.5
Neutral detergent fiber	58.7
Acid detergent fiber	8.2
Hemicellulose	50.5
Cellulose	5.8
Lignin	2.4
In vitro dry matter digestibility	97.6

From Bressani and Chon [14].

content varies from 4.60 to 12.17%, depending on the drying process used (table 1). As regards storing this material, Puleston [7] briefly describes an experiment carried out in 1968 in which beans, corn and corn tree seed were stored under the same conditions in a kind of silo found in Tikal, Guatemala [7]. These silos at at Tikal, called 'chultunes', are believed to have been used by the Mayas to store these and other staple foodstuffs. The study showed that after 9 weeks, the corn and beans were completely decomposed, while the corn tree seed was not. This may be attributed to certain physical or chemical properties that make the seed less perishable under the storage conditions used.

Chemical Composition

Bressani and Chon [14] report certain data regarding the composition of the corn tree, obtained at the Institute of Nutrition of Central America and Panama (INCAP), Guatemala City, Guatemala. The data obtained for the whole fruit (for animal feed purposes) are 12.3% of crude protein, 8% of water content, and 15.5% of ash (table 2). Table 1 shows the results of the analysis of proximal chemical composition of corn tree seed. In general, there are important variations between the results obtained: this may be attributed to the fact that the seed varies according to degree of ripening, season when cut, factors that influence its composition and quality [15]. These variations have not been controlled by using consistent sampling methods, which may identify the sources of variation that influence data reported. In spite of these results, it may be concluded that the seed is carbohydrate rich; values ranging from 39.6 to 74.6% of nitrogen-free extracts have been reported.

For ether extract, values from 0.86 to 2.76 (table 1) have been reported.

Crude fiber content varies from 2.4 to 8.9% (table 1); and total dietary fiber, varies from 16.6 to 23.6 mg [3].

The seed's caloric content, as found in pertinent reports, varies from 3.59 to 4.16 kcal/g [3]. This confirms the concept expressed by Pardo-Tejeda and Muñoz [9] that for this type of food, the corn tree seed has a high caloric content.

Puleston [2] found protein values for the seed varying from 11.4 to 13.4% of crude protein. These data are in the upper range of the reports made by other researchers, while data obtained at later dates at INCAP (1992) show crude protein values of 7.7–8.9% (table 1). These data compare favorably with wheat, corn and rice, the nutritional cornerstones of the main civilizations that have developed on the planet, whose average protein values are 9.3, 9.8 and 7.2%, respectively [13]. In many tropical regions, cultural considerations or ecologic conditions have contributed to the substitution of tubers for cereals. This is also seen at high altitudes. Such substitution makes it harder for the population to meet their nutritional needs, as most roots have a very low protein content; e.g. yams have 1.3% of crude protein and the potato 1.8% [13]. Thus, the use of corn tree as a partial substitute for corn in order to satisfy the nutritional requirements of populations living in the wet tropics (with shallow topsoil, low natural fertility, high alkalinity) where corn is not an efficient crop is a very good alternative. The Mayan culture must have known the importance of this plant's nutritive potential and its importance in the conservation of the area's forests.

Amino Acid Content and Protein Quality of the Corn Tree Seed

The essential amino acid content of corn tree seed, as reported by different institutions, is shown in table 3. From these data, it may be inferred that it is a high-quality protein.

Two amino acids are of special importance: lysine and tryptophan. These are limited nutrients in the common diets of Central America. Regarding lysine, values from 2.34 to 4.0% have been reported; for tryptophan, 1.2–2.3% in the corn tree seed [3, 9]. Even though there is not much information regarding sulfur-containing amino acids, Pardo-Tejeda and Muñoz [9] claim that the corn tree seed meets the amino acid levels suggested by the FAO/WHO (table 3). It is advisable to obtain more data on the sulfur-containing amino acid of the corn tree seed.

Asenjo [3] evaluated the biological response to protein quality of Wistar rats fed corn tree, using the net protein ratio method and apparent digestibility. Conclusions reached include that for the substitution levels used when adding corn tree to the diet, a smaller weight gain and protein intake was noted. The maximum net protein ratio value found for corn tree was 1.57, in contrast with 3.23 for casein.

Table 3. Essential amino acid content of the corn tree seed

Amino acid, %	INPI	ABL	FAO/WHO
Leucine	10.4	8.8	5.8
Valine	9.7	3.3	5.1
Arginine	5.1	4.0	4.9
Isoleucine	3.3	3.0	3.0
Phenylalanine	4.0	4.4	2.5
Lysine	2.3	4.0	2.3
Threonine	2.4	8.8	2.1
Tryptophan	2.3	1.2	1.4
Histidine	1.0	2.8	0.8
Methionine	0.7	–	–
Cystine	9.9	–	–

INPI, Mexico, 1976; ABL, 1976. From Pardo-Tejada and Muñoz [9].

The results of the apparent digestibility of dry matter for diets with different proportions of corn tree were high, going from 86.8% for the 40/60; corn/corn tree diet, to 94.3% for the 30/20: corn/corn tree diet. This is slightly lower than the digestibility of corn alone (96.3%). In this report, the net protein ratio results are not very encouraging; in studies developed with poultry and pigs, where marginal protein levels were not used [as they were by Asenjo, 3], the results are slightly more positive [12].

Development Potential

Tortillas and Bread

Food products made with corn flour may also be made with corn tree seed. In order to make tortillas, the seed is cooked in an alkaline medium; this is done in order to be able to later remove the skin by hand. Another alternative for peeling the seed is by roasting, which gives the seed an additional flavor, while making it easy to remove the skin. The seed is then washed, milled, kneaded and cooked in a process similar to that used for corn tortillas [16, 17].

In order to make bread, dough must be made from the seed using the roasting process. The dough is mixed in a no more than 1:1 ratio of corn tree flour to wheat flour, then baked as regular bread. When the amount of corn tree is higher than that described, product acceptability diminishes [6, 16, 17].

Before the final cooking process for tortillas or bread, the dough may be diluted with water and boiled; this beverage is known as 'atol'. Sugar, salt and/or cinnamon are usually added.

Animal Feed

Leaves. Yerena et al. [10] reported 44% of dry matter and 4.5% of nitrogen content for corn tree leaves, which is equivalent to slightly lower than 30% of crude protein. Also, voluntary consumption for cattle was 5.89 kg dry matter/100 kg live weight/day. These data show the high potential that this plant has for ruminant feed in this ecological zone.

Seed. The chemical composition of the seed, as reported by Pardo-Tejeda and Muñoz [9] and of the whole fruit [14] (table 1), indicates that both portions of the plant have high potential for ruminant feed. Based on these findings, research has been done on replacing sorghum with this product in poultry and pig feed [11, 12].

Lozano et al. [12], when replacing sorghum with corn tree seed at 0% and 50%, in chicken feed with 23% protein, found no significant difference in weight gain (646.5 vs. 587.0 g), respectively, in a 30-day trial. Nor was a difference found in feed conversion (1.79 vs. 2.11, respectively). Complete substitution (100%), however, did affect these parameters. The same authors were also unable to find a significant difference when replacing sorghum with corn tree seed in pig starter feed. At 0% and 30%, weight gains of 26.2 kg and 25.6 kg, respectively, were noted, as well as feed conversion of 3.93 and 4.07, respectively. When corn tree was substituted for sorghum by 60%, a negative effect was found on both parameters [12].

Forestal Resources

The most direct form of using corn tree is that of collection, processing, distribution and use of seeds already present; planting young trees is also necessary. In 1981, Peters and Pardo-Tejeda [8] estimated that in the State of Veracruz (Mexico), thousands of metric tons of seed could be collected annually. These could then be used as a human and animal food source. The National Institute of Biotic Research in Mexico concludes that the most important problem in using this resource to its fullest lies in marketing; however, if a product competes in cost and nutritive value, it will generate demand by the processing plants that make feed, which are always alert to the possibility of maximum benefit at minimum cost. On the other hand, if a product is acceptable to the consuming public and is a nutritive product, there seems to be no reason to avoid developing adequate marketing channels.

Considering the environmental factors, the planting of corn tree would contribute to the conservation of forestal resources in areas such as Petén in Guatemala, which although having been declared a valuable and protected area, is currently suffering the loss of many valuable trees, plants and animals.

Conclusions

The corn tree is still a widely distributed and abundant tropical resource.

Available literature indicates that the corn tree has a high potential for utilization as foodstuff and livestock forage.

Currently, the most direct use for the corn tree in the ongoing search for alternate food sources in Central America seems to be the use of its seeds for ruminant and monogastric fodder, and of seeds and leaves for ruminants.

As foodstuff or forage, corn tree seeds are rich in crude protein, particularly tryptophan and lysine, two of the most limited amino acids in the Mexican and Central American diet.

Increased utilization of corn tree as forage may contribute in preventing the destruction of rain forests. Tropical forests are presently being destroyed at an alarming rate, ironically to create new grazing and pasture areas. However, cattle are not the only reason rain forests are destroyed; cash crops and lumber play a role. Thus, lack of knowledge and technology also contribute to the destruction of the rain forest, with its ecological, economic, and food-producing potential.

References

1   Guzmán UAR: 'Conozcamos el Ramon (*Brosimum alicastrum* SW), Guatemala, Ministerio de Agricultura, Ganadería y Alimentación, Unidad de Comunicación Social, 1986.
2   Puleston PO: El Ramon como Base de la Dieta Alimenticia de los Antiguos Mayas de Tikal. Nuevos Datos sobre Subsistencia Alimenticia en el Maya Clásico). Guatemala, Instituto de Antropología e Historia de Guatemala II Epoca, 1979, vol 1(1), pp 55–69.
3   Asenjo Cabral C: Caracterización y usos de la semilla del arbol de 'ramon' *(Brosimum alicastrum).* Centro de Estudios Superiores en Nutrición y Ciencias de Alimentos (CESNA) Curso de postgrado en Ciencias y Tecnología de Alimentos, Universidad de San Carlos de Guatemala, Facultad de Ciencias Químicas y Farmacia, Instituto de Nutricíon de Centro América y Panamá (INCAP), 1992.
4   Lundell CL: The Vegetation of Petén. Washington, Carnegie Institution of Washington, 1937, pp 98–99, 107, 122, 141–144, 208.
5   Lambert JDH: Ramon and Maya ruins: An ecological, not an economic, relation. Science 1982;216: 298–299.
6   Melgar M, Aragón B, Méndez LF, Cuevas R: Utilización Integral del Arbol del Género Brosimum (Informe Final del Proyecto). Guatemala, Instituto de Nutrición de Centro América y Panamá (INCAP), 1987.
7   National Academy of Science: Underexploited Tropical Plants with Promising Economical Value, ed 7. Washington, National Academy Press, pp 114–118.

8   Peters CM, Pardo-Tejeda E: *Brosium alicastrum* (Moraceae): Uses and potential in México. Econ Bot 1981;36:166–175.

9   Pardo-Tejada, Muñoz CS: Ramon, capomo, ojite, ojoche, *Brosimum alicastrum:* Recursos Silvestre Tropical Desaprovechado. Veracruz, Instituto Nacional de Investigaciones sobre Recursos Bióticos. Cuadernos de Divulgación, 1974.

10   Yerena F, Ferreiro H, Elliot R, Godoy R, Preston T: Digestibilidad de ramon *(Brosimum alicastrum), Leucaena leucocephala,* pasto buffel *(Cenchus ciliane),* y pulpa y bagazo de Henequen (Agave Fourcroydes). Producción Animal. Tropical 1978;3:70–73.

11   Ramirez HJ: Valor energético de la semilla de ramon *(Brosimum alicastrum)* en dietas para aves. Técn Pecuaria (México) 1978;35:43–47.

12   Lozano O, Shimada A, Avila E: Valor alimenticio de las semillas del ramón *(Brosimum alicastrum).* México, Instituto Nacional de Investigaciones Pecuarias. Asociación Latinoamericana de Producción Animal (ALPA), 1978, vol 13.

13   WV-Leing WT, Flores M: Tabla de Composición de Alimentos para Uso en América Latina. Guatemala, INCAP, 1961.

14   Bressani R, Chon C: Composición Química, Fraccionamiento Celular y Digestibilidad In-Vitro de Frutas de Arboles Tropicales y de Leguminosas no Convencionales de Guatemala; in Memorias de II Congreso Nacional de la Carne y la Leche. AGSOGVA. Retalhuleu, December 1990.

15   Duckworth RB: Frutas y Verduras. Zaragoza, Editorial Acriba, 1968, chapt 1.

16   Mashkowski G: Usos del arbol de ramon *Brosimum alicastrum,* Guatemala; report, Justus-Liebig University, Giessen, 1988.

17   Pineda RG: El pan de harina de ramón (*Brosimum alicastrum* Swartz). Guatemala, Instituto Nacional Forestal, Departamento de estudios y proyectos, 1985.

M. Ortiz, Institute of Nutrition of Central America and Panama,
Pan American Health Organization INCAP/PAHO, Calzada Roosevelt Zona II,
Apartado Postal 1188, Guatemala C.A. (Guatemala)

Simopoulos AP (ed): Plants in Human Nutrition.
World Rev Nutr Diet. Basel, Karger, 1995, vol 77, pp 147–154

....

Hawthorn (Shan Zha) Drink and Its Lowering Effect on Blood Lipid Levels in Humans and Rats

J.D. Chen, Y.Z. Wu, Z.L. Tao, Z.M. Chen, X.P. Liu

Institute of Sports Medicine, Beijing Medical University, Beijing, China

Contents

Introduction 147
Materials and Methods 149
    Animal Experiment 149
    Human Study 149
    Indices 149
Results and Analysis of the Animal Experiment 150
Results and Analysis of the Human Study 153
Conclusions 154
References 154

Introduction

Hawthorn *(Crataegus pinnatifida)* is a wild, fruit-bearing tree growing in the mountains of China. Its berry tastes sour and sweet and is considered as both a fruit and a herb. Hawthorn is used not only for medicinal purposes, but is also sold on street corners as a popular food. It has been accepted and reported that hawthorn has some natural and effective components including haw flavone, triterpenoid, rutin, tartaric acid, citric acid, crataegolic acid, ester, glucoside, analytical lipid enzyme, carbohydrate and some saponins [1]. The Chinese generally consider it useful in reducing food stagnancy and blood stasis. As a medicine, it is used to treat hypercholesterolemia, angina pectoris, and hypertension, because its

图 411 山里红

图 410 野山楂

Fig. 1. Another species of haw (Shan Zha). From *Compilation of Chinese Herbs* [2].

Fig. 2. Wild hawthorn (Shan Zha). From *Compilation of Chinese Herbs* [2].

components possess effects of increasing the coronary artery blood flow rate, reducing blood cholesterol concentration and decreasing myocardial oxygen consumption [1]. The plant is represented in figures 1 and 2 [2].

Atherosclerosis is one of the most serious diseases endangering human health [3]. It has become the first cause of death. Hyperlipidemia is known as one of the most important risk factors of atherosclerosis. Development and utilization of the natural resource of hawthorn to reduce blood lipids will not only satisfy people's need for drinks from natural plant sources, but will also sustain health and provide economic benefits.

Using a natural plant, we adopted a low sugar content and fortified the drink with adequate amounts of vitamin C and zinc gluconate to enhance the remarkable antioxidative features of the plant. The effective components and nutrients of this hawthorn drink are listed in table 1 and fully illustrate the characteristics of the drink: low sugar, low sodium, high iron and high zinc content, and adequate amounts of vitamins B_1, B_2 and C.

In order to investigate the effects of the hawthorn drink on reducing body weight, body fat and blood lipids, we carried out studies on rats and humans in 1992.

Table 1. Nutrient composition of the hawthorn drink

Component	Amount/100 ml
Haw flavones	560 µg
Sugar (cane)	4–6 g
K	19 mg
Na	17 mg
Ca	7 mg
Mg	4 mg
Zn	278 µg
Cu	2 µg
Fe	250 µg
Vitamin B_1	106 µg
Vitamin B_2	390 µg
Vitamin C	28 mg

Materials and Methods

Animal Experiment

37 male weight-matched Sprague-Dawley rats (80–90 g) were divided into three groups as follows: (1) experimental group: ingesting hawthorn drink (n = 13); (2) control group: ingesting 8% sugar water (n = 11); (3) control group: ingesting tap water (n = 13). The animals were kept in individual cages and raised by complete nutrition feed for 2.5 months and then killed in pairs and batches.

Human Study

Subjects. 30 volunteers from the hospital staff, clinically diagnosed as hyperlipidemic, participated in the study. Among them 25 were females and 5 were males aged 51.1 ± 8.0 years. During the experimental period (1 month), the subjects kept their usual daily life, activity, and diet, but stopped taking medications. The subjects came to the lab and took the hawthorn drink twice a day (250 ml each time) for 1 month. Venous blood samples were obtained in the fasting state in the early morning at the beginning and at the end of the experimental period.

Indices

Body Weight of Rats. The rats were weighed once a week with a Tanita electronic balance (sensitivity: 0.2 g/100 g).

Body Fat of Rats. By weighing perirenal and peritesticular fat (data from our previous work showed that fat in these two sites is closely correlated with total body fat; $r = 0.85$, $p < 0.001$).

Food and Fluid Intake of Rats. These parameters were recorded every day and the data expressed as average quantity/week.

Blood Lipids for both Humans and Rats. Total cholesterol: cholesterol esterase and oxidase reagent kit; triglycerides: acetyl acetone spectrophotometry; high-density lipoprotein cholesterol (HDL-C): enzyme reagent kit; low-density lipoprotein cholesterol (LDL-C):

Table 2. Effects of hawthorn drink on the body weight of rats (g)

Experimental period	Group 1 n = 13	Group 2 n = 11	Group 3 n = 13
1st week	109 ± 13.5	109 ± 13.7	110 ± 12.5
5th week	292 ± 32.1	294 ± 28.6	309 ± 19.1
9th week	395 ± 31.8	416 ± 37.8	419 ± 28.9
10th week	403 ± 35.1[a]	428 ± 36.8	432 ± 30.2
Before sacrifice	385 ± 34.0[a, b]	419 ± 33.8	419 ± 30.9

[a] $p < 0.05$, group 1 (hawthorn) vs. group 3 (tap water).
[b] $p < 0.05$, group 1 vs. group 2 (sugar water).

enzyme reagent kit; Apo AI and apo B: immunological turbidimetry; malonic dialdehyde: TBA reaction by fluorospectrophotometry.

Statistics. Data are expressed as means ± SD. Comparisons between different groups of data obtained were calculated by Student's t test.

Results and Analysis of the Animal Experiment

Effects of Different Drinks on BW of Rats. The body weight of rats ingesting hawthorn tended to be lower than that of the two control groups during the first 5 weeks, but the differences were statistically nonsignificant. From the 6th to 9th week, body weight differences between the three groups of rats became clearer but they were still nonsignificant. By the 10th week, the rats ingesting the hawthorn drink showed significantly lower body weights and the weights were also significantly lower than in the control groups before sacrifice (table 2).

Effects of Different Drinks on Body Fat of Rats. The data show that the body fat of rats ingesting the hawthorn drink are significantly less than in the controls. There is no significant difference between rats ingesting sugar water or tap water (table 3). The results indicate that the hawthorn drink is effective in reducing the body fat of rats.

Effects of Hawthorn Drink on Serum Lipid Levels of Rats. The results show that the serum cholesterol levels of rats ingesting the hawthorn drink significantly lower than those of the two control groups ($p < 0.001$; table 4). The cholesterol levels of the rats ingesting the hawthorn drink were all in the normal range while the levels were above normal in some of the rats of the control groups. Changes in the levels of triglycerides after taking the different drinks showed the same pat-

Table 3. Effects of the hawthorn drink on body fat of rats

Groups	n	BW, g	PRF + PTF g/100 g BW
1	13	385 ± 34[a, b]	3.56 ± 0.63[a, b]
2	11	419 ± 34	4.32 ± 1.04
3	13	419 ± 31	4.19 ± 0.85

PRF = Perirenal fat; PTF = peritesticular fat.
[a] $p < 0.05$, group 1 (hawthorn) vs. group 3 (tap water).
[b] $p < 0.05$, group 1 vs. group 2 (sugar water).

Table 4. Effects of hawthorn drink on serum lipids of rats

Groups	n	Cholesterol mmol/l	Triglyceride mmol/l
1	13	1.48 ± 0.16[a]	1.32 ± 0.16[b]
2	11	1.82 ± 0.21	1.51 ± 0.20
3	13	2.03 ± 0.31[c]	1.42 ± 0.22

[a] $p < 0.01$, group 1 (hawthorn) vs. group 2 (sugar water).
[b] $p < 0.05$, group 1 vs. group 2.
[c] $p < 0.001$ group 1 vs. group 3 (tap water).

tern. These results indicate that the hawthorn drink is effective in preventing an increase in blood cholesterol and triglycerides above normal levels.

Effects of the Different Drinks on HDL-C and LDL-C Levels. HDL-C levels were significantly greater in rats ingesting the hawthorn drink than those of the rats taking sugar water but less than those of the rats drinking tap water. LDL-C levels of rats ingesting the hawthorn drink were the lowest among the three groups of rats, and the difference was highly significant ($p < 0.001$; table 5). The result shows that the hawthorn drink helps to reduce the LDL-C levels, and thus to prevent atherosclerosis [4].

Effects of Different Drinks on Food and Fluid Intakes. Data on the average food intake of rats show that it was the least in rats drinking sugar water, and it was significantly less as compared with the data of rats drinking tap water. The food intake of rats ingesting the hawthorn drink ranged between the data of rats

Groups	n	HDL-C mmol/l	LDL-C mmol/l
1	13	0.93 ± 0.07^a	0.26 ± 0.15^a
2	11	0.76 ± 0.11^b	0.59 ± 0.10
3	13	1.10 ± 0.20^c	0.53 ± 0.19^d

Table 5. Effects of the hawthorn drink on lipoprotein levels of rats

[a] $p < 0.001$, group 1 (hawthorn) vs. group 2 (sugar water).
[b] $p < 0.001$, group 2 vs. group 3 (tap water).
[c] $p < 0.01$, group 1 vs. group 3.
[d] $p < 0.001$, group 1 vs. group 3.

Experimental period	Group 1 (n = 13)	Group 2 (n = 11)	Group 3 (n = 11)
2nd week	16.5 ± 1.5^a	14.2 ± 0.9	15.0 ± 2.4
5th week	20.0 ± 1.9^a	17.6 ± 1.4	22.6 ± 2.1^b
9th week	21.0 ± 1.4^c	17.3 ± 1.5^d	24.1 ± 1.2^e
10th week	19.3 ± 1.5^c	17.3 ± 3.0^d	22.4 ± 1.6^e

Table 6. Average feed amount of rats with different fluid intakes (g)

[a] $p < 0.01$, group 1 (hawthorn) vs. group 2 (sugar water).
[b] $p < 0.05$, group 1 vs. group 3 (tap water).
[c] $p < 0.001$, group 1 vs. group 2.
[d] $p < 0.001$, group 2 vs. group 3.
[e] $p < 0.001$, group 1 vs. group 3.

Experimental period	Group 1 (n = 13)	Group 2 (n = 11)	Group 3 (n = 11)
2nd week	32.0 ± 7.2	25.7 ± 5.1	24.9 ± 2.3^a
5th week	37.1 ± 2.0^b	56.7 ± 4.5^c	37.1 ± 3.2
9th week	45.6 ± 4.6^b	71.8 ± 9.4^c	52.0 ± 3.9^d
10th week	51.2 ± 3.9^b	71.9 ± 5.6^c	52.7 ± 3.6

Table 7. Average fluid intakes of rats with different drinks (ml)

[a] $p < 0.05$, group 1 (hawthorn) vs. group 3 (tap water).
[b] $p < 0.001$, group 1 vs. group 2 (sugar water).
[c] $p < 0.01$, group 2 vs. group 3.
[d] $p < 0.01$, group 1 vs. group 3.

Table 8. Effects of hawthorn drink on blood lipid, lipoprotein, and apolipoprotein levels of hyperlipidemia patient (n = 30)

Indices	Before treatment	After treatment	p
Serum cholesterol, mmol/l	7.31±1.04	6.19±1.56	<0.001
Serum triglycerides, mmol/l	1.93±0.92	1.75±0.96	<0.05
HDL-C, mmol/l	1.07±0.18	1.12±0.13	>0.05
LDL-C, mmol/l	3.95±1.14	3.54±0.96	<0.001
apo-AI, mg/dl	125±23	124±20	>0.05
apo-B, mg/dl	95±16	88±19	<0.05
Serum MDA, nmol/ml	3.2±0.9	2.5±0.4	<0.001

drinking sugar or tap water, although it was significantly less than that of the rats drinking tap water at the 5th, 6th, 9th and 10th week. The intakes were basically the same in these two groups during most of the experimental period (table 6).

Data on the fluid intake show that the amount was the maximum in rats drinking sugar water from the 3rd week to the 10th week, and it was significantly larger than that of rats drinking tap water or the hawthorn drink. The fluid intake of rats ingesting the hawthorn drink was similar to that of rats drinking tap water for most of the experimental period (table 7).

These results indicate that the feed intake was about one third less and the fluid intake one third more in rats drinking sugar water compared with rats drinking tap water; the average body weight of the rats of these two groups were the same. The feed intake of rats ingesting hawthorn was slightly larger, but fluid intake was far less as compared with the rats taking sugar water, thus, the body weight, body fat and blood lipids of the rats ingesting the hawthorn drink were the lowest.

Results and Analysis of the Human Study

Effects of the Hawthorn Drink on Serum Cholesterol and Triglyceride Levels of Hyperlipidemic Patients. The average serum cholesterol level of the subjects was 7.30 ± 1.04 mmol/l and was reduced to 6.19 ± 1.56 mmol/l after ingesting the hawthorn drink for 1 month; the difference was highly significant ($p < 0.001$). The cholesterol levels of 17 subjects (56.7% of the total) decreased to <6.24 mmol/l. Serum triglyceride levels decreased from 1.93 ± 0.92 to 1.75 ± 0.96 mmol/l (table 8).

The results showed that the hawthorn drink is effective in reducing serum cholesterol and triglyceride levels of hyperlipidemic patients.

Effects of the Hawthorn Drink on Lipoprotein and Apolipoprotein Levels of Hyperlipidemic Patients. The results were similar to those of the animal experiment. Hawthorn was effective in reducing LDL-C levels, but not effective in raising the HDL-C levels (table 8). Similarly, the hawthorn drink was not effective in raising the apo-AI levels, but very effective in reducing the apo-B levels. Apo-B levels (apo-B is the main component of LDL-C) decreased in 70% of the subjects.

Effects of the Hawthorn Drink on Lipid Peroxidate Malonic Dialdehyde (MDA) Levels. Average serum MDA levels of subjects decreased from 3.2 ± 0.9 to 2.5 ± 0.4 nmol/ml after ingesting the hawthorn drink for 1 month; the difference was significant ($p < 0.001$) (table 8). Moreover, 10 of the 11 subjects with high MDA levels all reduced to <3.3 nmol/ml; this result showed that the hawthorn drink possesses strong antioxidative effects because it contains multiple antioxidative components.

Conclusions

The hawthorn drink was effective in reducing body weight, body fat, serum cholesterol, triglyceride and LDL-C levels of rats.

Serum cholesterol, triglycerides, LDL-C and apo-B levels of hyperlipidemic patients were significantly reduced after ingesting the hawthorn drink for 1 month.

The hawthorn drink significantly reduces MDA and blood lipid levels of hyperlipidemic patients; these effects are very likely related to its multiple effective antioxidative components.

References

1 Huang KC: Shan Zha, *Crataegus pinnatifida;* in The Pharmacology of Chinese Herbs. Boca Raton, CRC Press, 1993, pp 101.
2 Compilation of Chinese Herbs: Compilation of the Nation's Herbs. People's Publishing House, Beijing, 1977, p 103, fig. 410, 411.
3 Chen JS: Changes of disease model and the countermove direction of prevention and treatment. J Chinese Prevent Med 1990;24:290–293.
4 Parthasarathy S, Rnikin SM: Role of oxidized low density lipoprotein in atherosclerosis. Prog Lipid Res 1992;31:127–143.

J.D. Chen, Institute of Sports Medicine, Beijing Medical University, Beijing 100083 (China)

Subject Index

Acetylcoenzyme A carboxylase 9
  silicon 14
Aleurone layer 91–94, 103
Algae
  fatty acid composition 9–14
  nutritional value 34–44
Algal oils 4–6, 23
Alkaloids, lupin 84, 85
Amaranthus
  hypochondriacus 64
  retroflexus 64
Amino acids
  corn tree seed 142, 143
  lupin 78, 79
  nopal 119, 121
  seeds 52
  spirulina 35–37
Amylopectin, barley 96
Amylose, barley 92, 93, 96
Animal feed, *see* Fodder, corn tree seed
Antioxidants 154
  purslane 48, 56–58, 60–63
Apolipoproteins 154
  hawthorn drink 154
Arabidopsis thaliana 24
Arabinoxylans 96
Arachidonic acid 2, 3, 7
  algae 10
  Porphyridium cruentum 17–19
Ascorbate 63
Ascorbate-glutathione cycle 63, 64

Ascorbic acid, purslane 57, 60
Atherosclerosis 148
Atol beverage 144

Bacillariophyceae, fatty acid composition
  14
Barley
  amylopectin 96
  amylose 96
  bran content 102
  carbohydrate content 95–98
  cellulose content 96, 97
  cholesterol metabolism 98–103
  flour 102
  foods containing 89–105
  genetic composition 91–93
  germ 95
  grain 93, 94
  hull-less 104
  hypocholesterolemic activity 98
  lignin 96, 97
  linoleic acid 94, 95
  α-linolenic acid 94, 95
  lipid content 94, 95
  lipid metabolism 102, 103
  lysine content 93
  malt 105
  milling 103–105
  mineral content 95
  nutritional value 93–98
  oil 100, 102

Barley (continued)
  pearled 91
  protein content 93
  protein efficiency ratio 93
  starch content 95, 96
  vitamin content 95
Beta vulgaris cicla 64
Bile acids 102, 103
Body fat, hawthorn drink 150, 151
Body weight, hawthorn drink 150, 151
Bran 98
  plasma lipids 100, 101
Bread, corn tree seed 143, 144
Brewer's grain 101, 102
Brosimum alicastrum, see Corn tree
B vitamins 41, 42
  spirulina 35

Calcium, nopal 123
'Camus ensara' 113
Cancer, fatty acids 2
Candida lipolytica 35
Carbohydrates
  barley 95–98
  nopal 123, 124
  spirulina 34
Cardiovascular disorders 2
Carminic acid 129, 130
β-Carotene 41, 42
  nopal 122
  purslane 57, 58, 61
  spirulina 35
Caryopsis 91, 92
Catalase 61
Cell concentration, *Porphyridium
  cruentum* 18
Cellulose, barley 96, 97
Chenopodium album 64
Chlamydomonas reinhardtii 24
Chlorella vulgaris 24
Chlorophyceae 13
Chloroplasts, purslane 63
Cholesterol
  hawthorn drink 150, 151, 153, 154
  metabolism, effect by barley 98–103
  nopal 126
Chroomonas salina 9

Chrysophyceae, fatty acid composition 5
Citrus pectin 70
Cladodes 109, 110
  hypoglycemic effect 125
Clones, herbicide-resistant 19–21
Cobalamine, spirulina 42
'Cochineal insect' 116, 117, 129, 130
'Colonche' 110, 131
Corn tree 135, 136
  fruit 141
  morphology 136, 137
  nutritional value 140–143
  production 139, 140
  seed 140–142
    amino acid content 142, 143
    protein content 142
    uses 143–145
  uses 138, 139
Coronary artery disease 48, 49
  fatty acids 2
Crassulacean acid metabolism 67
Crypthecodinium cohnii 9
Cyclotella
  cryptica 14
  meneghiniana 12
Cytofluorometry 25

Desaturase activity 61
Diabetes, treatment by nopal 125–127
Diatoms
  fatty acid composition 5
  light intensity 11, 12
Dietary fiber, *see* Fiber, dietary
Digalactosyldiacylglycerol 41
  purslane 56
Dinophyceae, fatty acid composition 5,
  14
Docosahexaenoic acid 2–4, 9
  biosynthesis 6–9
  distribution 14
  heterotrophic production 22
  Isochrysis galbana 19
  light 12
  purslane 55
Dopa, purslane 68
Dopamine, purslane 68

Egg white-pectin complex 70
Eicosapentaenoic acid 2–4
  algae 10, 15–19
  algal triglycerides 21, 22
  biosynthesis 6–9
  distribution 14
  herbicides 19–21
  heterotrophic production 22
  light 12, 13
  nitrogen 13
  phototrophic production 15–19
  Porphyridium cruentum 17–19
  purification 23
  purslane 55
  temperature 11
Erucic acid, lupin oils 79
Euglena gracilis 8, 10
Eustigmatophyceae, fatty acid composition 5, 14

Fat, spirulina 34
Fatty acids
  composition 9–14, 17–19
  coronary vascular diseases 2
  desaturation 7–9, 20, 21
  elongation 7–9
  lupin 78–80
  microalgae 1–25
  nitrogen 13
  overproduction 20, 21
  phototrophic production 15–19
  plants 49
  purslane 54–56, 60–63
  silicon 14
  spirulina 38–41
Fiber
  dietary 90–92, 96, 97
    barley 96, 97, 99, 102–104
    coronary artery disease 49
    nopal 119, 120, 123
    prickly pear 121
  insoluble 96, 97, 123
'Figue de barbarie' 113
Fish oil, ω3 fatty acid composition 4
Fluid intake, hawthorn drink 152, 153
Fodder, corn tree 138, 139, 144

Food intake, hawthorn drink 151–153
Forage, corn trees 144, 145

Galactolipids 40
  biosynthesis 61
  purslane 55, 56
β-Glucan 98, 99, 104
Glucose serum level, effect by nopal 125, 126
Glutathione, purslane 58–61
Glutathione-linked disulfides 60
Glycolipids, spirulina 38–41
Gracilaria verrucosa 6, 7
Growth temperature, fatty acid composition 10, 11

Hawthorn 147, 148
  drink 148, 149
    antioxidative effects 154
    fluid intake 152, 153
    food intake 151–153
    hypolipidemic effect, testing 149–154
    lipoproteins 151–154
    malonic dialdehyde 154
Herbicides 24
  effect on fatty acid composition 19–21
High-density lipoproteins, hawthorn drink 151–154
Hypercholesterolemia, barley 100
Hyperlipidemia 148
  hawthorn drink 153, 154
Hypocholesterolemia, pectin ingestion 70
Hypolipidemia, pectin ingestion 70

Indian fig 112
Infant formulae, fatty acids 3
Insoluble fiber, *see* Fiber, insoluble
Insulin serum level, effect by nopal 125, 126
Iron, spirulina 42
Isochrysis galbana, fatty acid composition 19

Leaf tissue, nutritional value 60–63
Leukotriene eicosanoids 2, 3

Light
fatty acid composition 11, 12
Nannochloropsis sp. 12, 13
Porphyridium cruentum 19
Light-dark cycles 12
Lignin, barley 96, 97
Linoleic acid 2, 3, 38, 79, 94, 95
α-Linolenic acid 2, 3, 38
barley 94, 95, 98
biosynthesis 6
cholesterol metabolism 100,
101
lupin 79
purslane 48, 54, 55, 61, 62
γ-Linolenic acid 39, 40
Linseed oil 10
Lipids
algae 10
barley 94, 95, 102, 103
hawthorn drink 150–154
polar, purslane 55
spirulina 38–41
nutritional value 40, 41
Low-density lipoproteins 49
hawthorn drink 151–154
nopal 126
Lupin
agricultural production 85, 86
alkaloids 84, 85
amino acid content 78, 79
fatty acid content 78–80
flour 82, 83
hulls 84
nutritional value 77–86
oil content 78, 79
part of human diet 82–86
protein
isolates 84
quality 80, 81
toxicity 8, 84
Lupinosis 84
Lupinus
albus 75–78
protein quality 81
angustifolius 77, 78, 81
limifolius 77
luteus 77, 78

mutabilis 77–79, 81
pilosus 77
Lysine
barley 93
corn tree seed 142
spirulina 36, 37

Malonic dialdehyde, hawthorn drink 154
Mercury, spirulina 42
Methionine
lupin 80, 81
spirulina 36, 37
Microalgae, fatty acids 1–25
Minerals
barley 95
nopal 123
spirulina 42, 43
MK8620 22
MK8805 22
MK8908 22
Monodus subterraneus 8, 10
eicosapentaenoic acid production 16
Monogalactosyldiacylglycerol 6, 7, 38,
41, 55, 56
Mortierella 4
Muscle relaxation, purslane 68, 69

Nannochloropsis oculata 8, 9
eicosapentaenoic acid production 15, 16
National Rural Dietary Food Survey 130, 131
Navicula saprophilla 13
Niacin, spirulina 42
Nile red 25
Nitrogen
balance
lupin consumption 81, 82
spirulina 37
deposition 37
digestibility 37
fatty acid composition 13
Porphyridium cruentum 19
'Nocheztli' 117
Nopal 109–115
amino acid content 121, 122
antidiabetic effect 125–127
carbohydrates 123, 124
circulating cholesterol 126

consumption 130–132
cultural issues 116–118
dye production 129, 130
environmental aspects 115, 116
low-density lipoproteins 126
medicinal properties 125–127
mineral content 123
morphology 118, 119
nutritional value 119–124
organoleptic qualities 124
peptic acid disease 126
pollination 115
production 127–130
protein content 119–122
vitamin content 122, 123
Noradrenaline, purslane 67, 68
Nucleic acid, spirulina 34, 35

Oat foods, lowering of cholesterol 100
Ochromonas danica 7
Oil content, lupin 78, 79
Oleic acid 94, 95
lupin 79
Opuntia
ficus-indica 112, 115
megacantha 112
robusta 115
streptacantha 115
stricta 115
tormentosa 115
Oryzanol 101
Oxalic acid, purslane 51

Palea 91
Palmitic acid 94, 95
Pearling, barley 91, 103
Pectin
purslane 49, 69, 70
reduction of platelet aggregation 70
Peptic acid disease, nopal 126
Phaeodactylum tricornutum 7, 14
eicosapentaenoic acid production 15
growth temperature 10
Phomopsins 84
Phomopsis leptostromiformis 84
Phosphatidylcholine 6
purslane 56

Phosphatidylethanolamine, purslane 56
Phosphatidylglycerol 38, 41
purslane 56
Phospholipids
purslane 55, 56
spirulina 38
Phototrophic algae 15–19
Phytate 95
Plants, C_4 metabolism 67
Platelet aggregation, pectin 70
Polysaccharides, spirulina 34
Polyunsaturated fatty acids 2–25, 38, 48,
60–63; *see also* specific types
Porphyridium cruentum 7–10
eicosapentaenoic acid 17–19, 23
fatty acid composition 20, 21
light 11
nitrogen 13
Portulaca oleracea 48, 64
origin 50
Potassium
nopal 123
purslane 67
Potherb 53
Prasinophyceae, fatty acid composition 5
Prickly pear 110, 114, 118, 119
cheese 110, 129, 131
nutritional value 120
consumption 130, 131
dietary fiber 121
juice 131
nutritional value 120, 121, 122
production 128, 129
protein content 121, 122
Prostaglandin 2
Protein
corn tree seed 142
digestibility 37, 81
efficiency ratio (PER)
barley 93
lupin 80, 81
spirulina 37
lupin 77, 78
nopal 119–122
prickly pear 121, 122
spirulina 34–38
Provitamin B, *see* β-Carotene

Prymnesiophyceae, fatty acid composition 5
Purslane
    agricultural potential 64–67
    amino acid composition 52, 53, 65
    antioxidant content 56, 57
    dopa content 68
    dopamine content 68
    fatty acid content 49, 54–56
    food technology 69, 70
    geographic distribution 50
    glutathione 58–60
    growth chamber studies 60–63
    medicinal uses 67–69
    muscle relaxation 68, 69
    noradrenaline content 67, 68
    nutritional value 47–50, 52, 54–63
    polar lipid content 55
    potassium content 67
    protein content 65
    spermatogenesis impairment 69

Quinolizidine 84, 85

Red algae, fatty acid composition 6, 7, 14
Rhodella, fatty acid composition 14
Rhodophyceae, fatty acid composition 5, 14

Salinity 10
SAN 9785 8, 20, 21
Seeds, amino acid composition 52
Shan Zha, see Hawthorn
Silicon, fatty acid composition 14
Sodium chloride concentration 10
Sorghum 144
Spermatogenesis impairment, purslane 69
Spinach, nutritional value 55, 57
Spirulina
    amino acid composition 35–37
    chemical composition 34–43
    history 33
    human food 43

lipid content 38–41
mineral content 42, 43
nutritional value 32–44
production 33, 34
toxicity 42, 43
vitamin content 41, 42
Spirulina
    maxima 33, 36
        fatty acid distribution 41
    platensis 33, 36, 39
        fatty acid distribution 41
Starch, barley 95, 96
Sulfolipids
    purslane 56
    spirulina 40, 41
Sulfoquinovosyldiacylglycerol 38, 41
Sweet lupins, see Lupin

Temperature and fatty acid composition 10, 11
Tenochtli 112
Testa 77
Tissue inflammation 2
α-Tocopherol 63
    purslane 57, 60–63
d-α-Tocotrienol 95, 98
Tocotrienols, cholesterol metabolism 100, 101
Tortillas, corn tree seed 143, 144
Transacylation 9
Triglycerides 9
    algae 10, 21, 22
        ω3 fatty acids 2–4, 15–19, 38
    hawthorn drink 150, 151, 153, 154
    nitrogen 13
Tryptophan, corn tree seed 142

Vitamins
    barley 95
    nopal 122, 123
    spirulina 35, 41–43

Waxy starch, barley 92